Math Mammoth
Grade 1-A Worktext

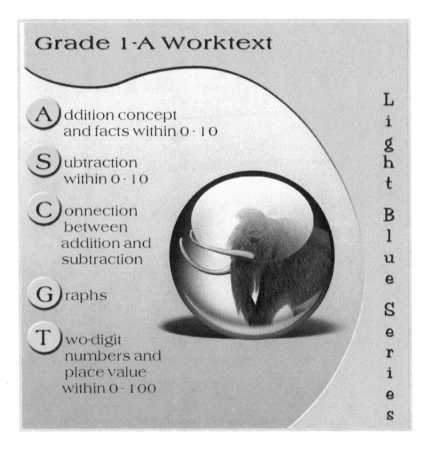

Grade 1-A Worktext

A ddition concept and facts within 0 - 10

S ubtraction within 0 - 10

C onnection between addition and subtraction

G raphs

T wo-digit numbers and place value within 0 - 100

Light Blue Series

By Maria Miller

Contents

Chapter 2: Subtraction Within 0-10

Chapter 3: Place Value Within 0-100

Foreword

Math Mammoth Grade 1 comprises a complete math curriculum for the first grade mathematics studies. The curriculum meets and exceeds the Common Core standards.

The main areas of study for first grade are:

1. The concepts of addition and subtraction, and strategies for addition and subtraction facts;

2. Developing understanding of place value up to 100;

3. Developing understanding and some basic strategies for two-digit addition and subtraction.

Additional topics we study in the first grade are telling time (whole and half hours), geometric shapes, measurement, and counting coins.

This book, 1-A, covers the concepts of addition and subtraction (chapters 1 and 2) and place value with two-digit numbers (chapter 3). The book 1-B covers strategies for addition and subtraction facts, clock, shapes and measuring, adding and subtracting two-digit numbers, and counting coins.

Some important points to keep in mind when using the curriculum:

- These two books (parts A and B) are like a "framework", but you still have a lot of liberty in planning your child's studies. While addition and subtraction topics are best studied in the order they are presented, feel free to go through the sections on shapes, measurement, clock, and money in any order you like.

  This is especially advisable if your child is either "stuck" or is perhaps getting bored with some particular topic. Sometimes the concept the child was stuck on can become clear after a break from the topic.

- Math Mammoth is mastery-based, which means it concentrates on a few major topics at a time, in order to study them in depth. However, you can still use it in a *spiral* manner, if you prefer. Simply have your child study in 2-3 chapters simultaneously. This type of flexible use of the curriculum enables you to truly individualize the instruction for your child.

- Don't automatically assign all the exercises. Use your judgment, trying to assign just enough for your child's needs. You can use the skipped exercises later for review. For most children, I recommend to start out by assigning about half of the available exercises. Adjust as necessary.

- For review, the curriculum includes a worksheet maker (Internet access required), mixed review lessons, additional cumulative review lessons, and the word problems continually require usage of past concepts. Please see more information about review (and other topics) in the FAQ at https://www.mathmammoth.com/faq-lightblue.php

I heartily recommend that you view the full user guide for your grade level, available at https://www.mathmammoth.com/userguides/

And lastly, you can find free videos matched to the curriculum at https://www.mathmammoth.com/videos/

I wish you success in teaching math!

Maria Miller, the author

Chapter 0: Kindergarten Math Review
Introduction

This chapter is optional and can be used to review the most important concepts of kindergarten math:

- writing the numerals 0 to 9;

- counting up to 20;

- position words, color words, and some shapes (circle, triangle, square)

- simple patterns

The Lessons in Chapter 0

Equal Amounts; Same and Different

1. Write an X for each thing in the other box.

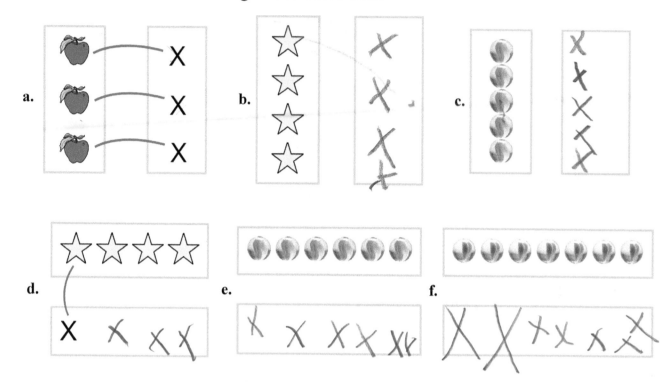

2. Color the shapes that are the same as the first shape.

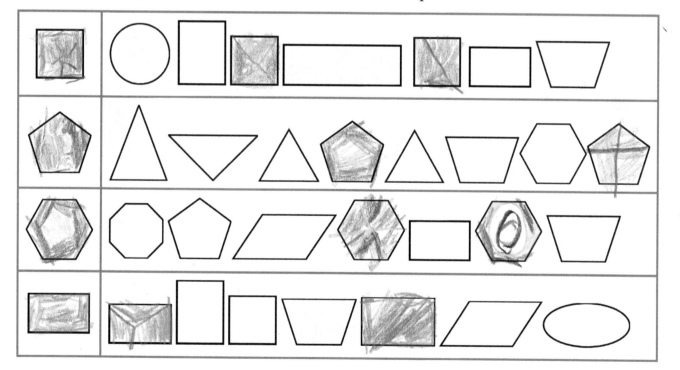

Writing Numbers

1. Write the number.

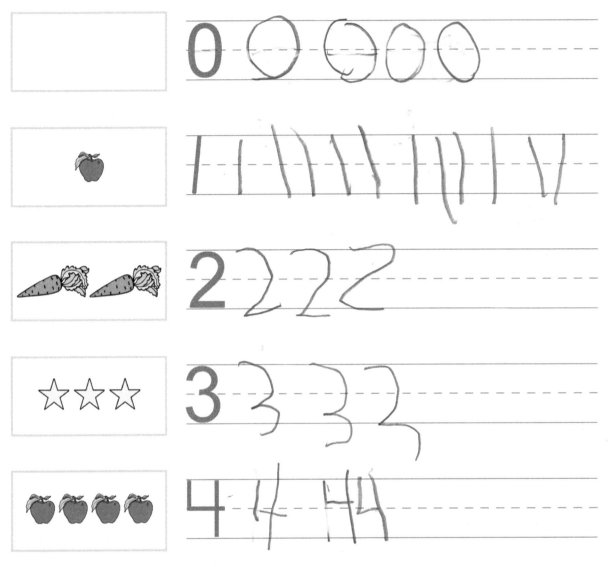

2. Count and write the number.

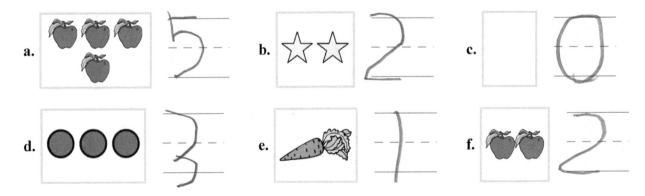

3. Write the number.

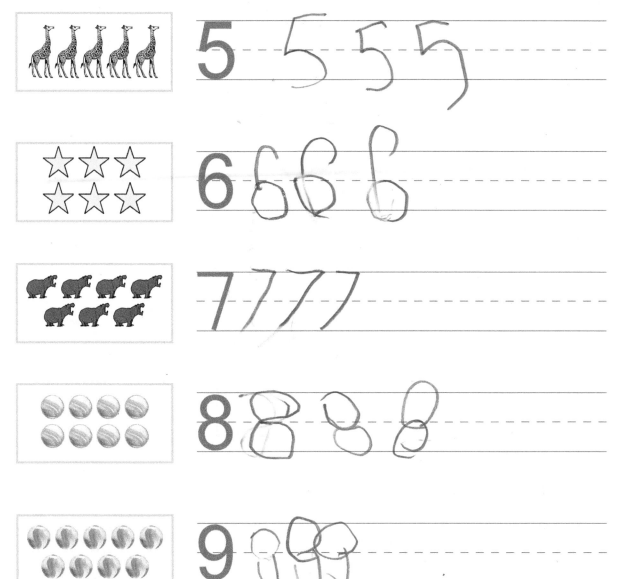

4. Count and write the number.

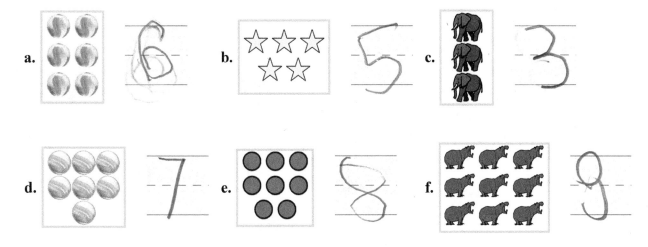

a.

b.

c.

d.

e.

f.

Counting

1. Count. Write the number in the box.

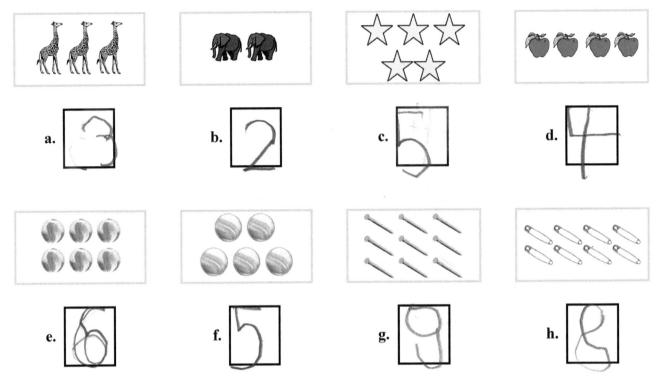

a. 3

b. 2

c. 5

d. 4

e. 6

f. 5

g. 9

h. 8

2. Count. Write the number. Then circle the number that is MORE.

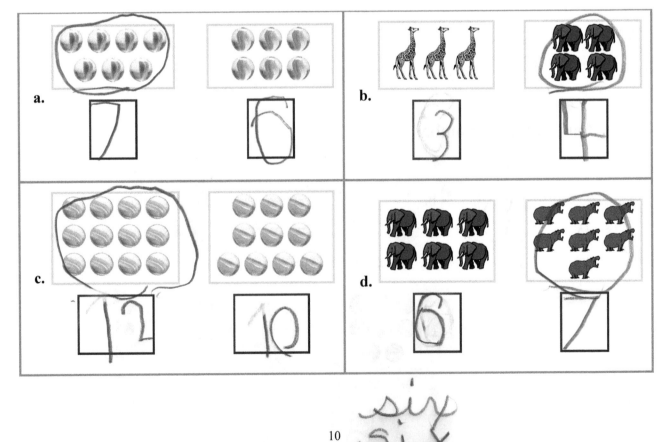

a. 7   6

b. 3   4

c. 12   10

d. 6   7

six
six

3. Write the missing number below the number line.

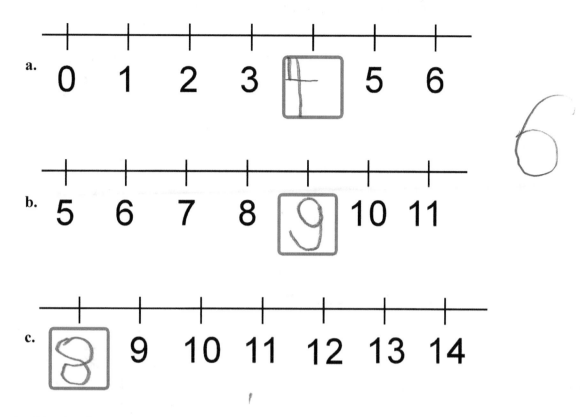

a. 0 1 2 3 [4] 5 6

6

b. 5 6 7 8 [9] 10 11

c. [8] 9 10 11 12 13 14

4. Circle the group that has more things. Then count ALL (both groups).
   Write the number in the box below.

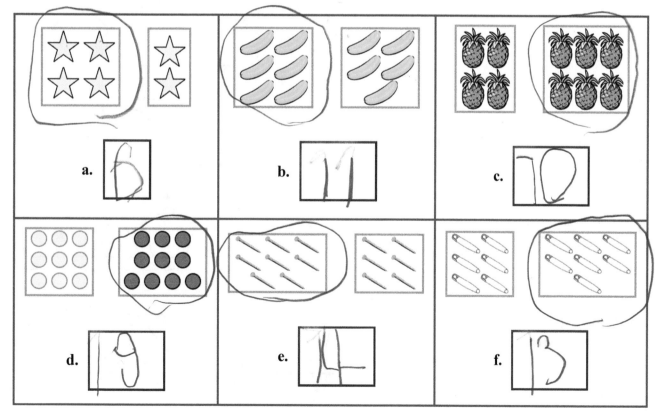

a. [6]

b. [11]

c. [12]

d. [9]

e. [14]

f. [13]

Position Words, Colors, and Shapes

1. **a.** Color the top shape RED.   **b.** Color the bottom shape BLUE.   **c.** Color the middle shape YELLOW.

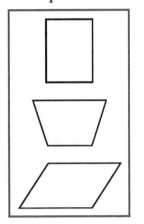 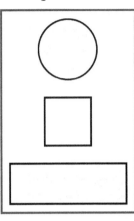 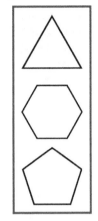

2. **a.** Color the shape on the right GREEN.

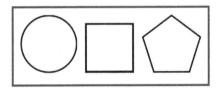

b. Color the shape in the middle BLUE.

c. Color the shape on the left YELLOW.

d. Color the two shapes on the right ORANGE.

e. Color the two shapes on the left PURPLE.

3. **a.** Color the bottom two shapes GRAY. **b.** Color the middle two shapes BROWN. **c.** Color the top two shapes BLACK.

4. **a.** Color all the circles GREEN.

b. Color all the triangles ORANGE.

c. Color all the squares PURPLE.

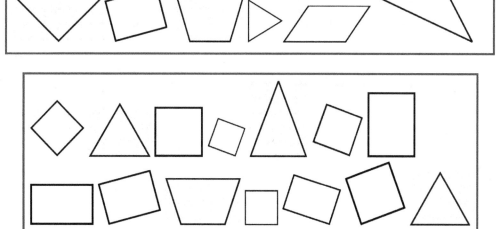

Patterns

1. Draw (in the correct color) the next shape in each pattern.

(P = purple, Y = yellow, B = blue, O = orange, R = red, B = blue)

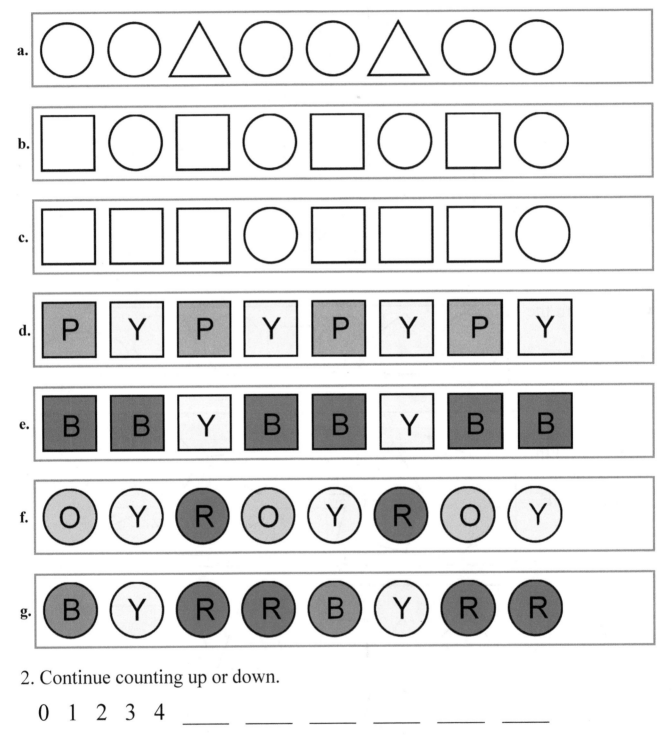

a.

b.

c.

d. P Y P Y P Y P Y

e. B B Y B B Y B B

f. O Y R O Y R O Y

g. B Y R R B Y R R

2. Continue counting up or down.

0 1 2 3 4 ____ ____ ____ ____ ____ ____

10 9 8 7 6 ____ ____ ____ ____ ____ ____

Chapter 1: Addition Within 0-10
Introduction

The first chapter of *Math Mammoth Grade 1-A* concentrates on the concept of addition and addition facts within 0-10.

Keep in mind that the specific lessons in the chapter can take several days to finish. They are not "daily lessons." Instead, use the general guideline that first graders should finish 1-2 pages daily or 7-9 pages a week. Please see the user guide at https://www.mathmammoth.com/userguides/ for more guidance on pacing the curriculum.

This chapter starts out with very easy and visual addition problems within 0-5, where children can simply count the objects to add. You can easily adapt these early lessons to be done with manipulatives (concrete objects such as blocks, beads, *etc.*).

If the student does not know the symbols " + " and " = " yet, you can introduce them *orally* at first. Use blocks or other objects to make addition problems and say: "Three blocks and four blocks makes seven blocks. Three blocks *plus* four blocks *equals* seven blocks." Then ask the child to make an addition with the objects, using those words. Play like that until the child can use the words "plus" and "equals" in his or her own speech. This will also make it easier to learn to use the written symbols.

In the lesson *Which Is More?*, the symbols " < " and " > " are introduced as being like a "hungry alligator's mouth." In this lesson, children only compare numbers, such as 5 < 7. In later lessons, children will also learn to compare expressions, such as 2 + 3 < 4 + 4.

Soon we introduce "missing addend" problems, or problems such as 1 + ___ = 5. First, we use pictures, and then gradually use only symbols. These problems are very important, as they lead the child to learn the connection between addition and subtraction.

Children might confuse the missing addend problem 1 + ___ = 5 with 1 + 5 = ___ . To help the child see the difference, you can word these problems like this: "One and how many more make five?"

You can model missing addend problems by drawing. In our example problem (1 + ___ = 5), the teacher would first draw one ball and then tell the student, "We need a total of five balls. Draw more balls until there are five of them." The number of balls that the child needs to draw in order to make five is the number that goes on the empty line. So you can say, "First there was one ball, then you needed to add (draw) some more to make 5. How many more did you draw?"

Then we come to the lesson *Sums with 5*. It practices the number combinations that add up to 5, which are 0 and 5, 1 and 4, and 2 and 3. After that we study sums with 6, sums with 7, and so on. The goal of these lessons is to help the child to memorize addition facts within 10. However, your child does not need to fully memorize them yet. All of these lessons are building toward that goal, but the final mastery of addition facts does not have to happen this early in first grade.

My approach to memorizing the basic addition facts within 10 is many-fold:

1. Structured drill, such as is used in the lessons *Sums with 5*, *Sums with 6*, and so on, are not random drills, because they use the pattern or the structure in the facts. This will connect the facts to a context, and help the child to better understand them on a conceptual level, instead of merely memorizing them at random. In each of these lessons, the child learns the number combinations that add up to the specific number. This understanding is the basis for the drills.

2. Using addition facts in games, in math problems, in every day life, or anywhere else are especially useful because most children like to play games.

3. Random drilling may also be used, sparingly, as one tool among others.

4. Memory helpers, can be silly mnemonics or writing math facts on a poster and hanging it on the wall. Not all children need these, but feel free to use them if you like.

These same addition facts are studied again in the following chapter about subtraction. They are also used constantly in all later math work. I recommend that children become fluent with addition facts within 0-10 by the end of first grade (as is also mentioned in the Common Core Standards).

Another important thread running through the chapter is to develop children's understanding of the symbols +, <, and > . Children need to get used to equations like $9 = 5 + 4$ and inequalities like $2 < 5 + 4$. They need to understand the equation $2 + \_\_\_ = 6$ correctly as an unknown addend problem, and not as the addition problem $2 + 6$. We need to prevent the misconception of the equal sign being an "operator," as if it means that you need to add/subtract/multiply/divide, or "operate" on the numbers in the equation. A child with this misconception will treat the equation $9 = \_\_ + 4$ as an addition problem $9 + 4$.

The chapter involves a lesson about addition on a number line, which is an important way to model addition. Children also encounter addition tables, number patterns, word problems, and get used to a symbol for the unknown number (such as in $\boxed{} + 5 = 10$). So, while it may look on the surface that all we do is add small numbers, actually, a lot happens in this chapter!

Please also see the following page for a few games that I recommend while studying this chapter. Games are important at this level, as they help children to practice the addition facts and also make math fun. Don't forget to check out the free videos matched to the curriculum at https://www.mathmammoth.com/videos/.

The Lessons in Chapter 1

Games for Addition and Subtraction Facts

10 Out (or *5 Out* or *6 Out, etc.*)

You need: Lots of number cards with numbers 1-10, such as regular playing cards (without the face cards), or any other cards that have numbers on them.

Rules: Deal seven cards to each player. Place the rest face down in a pile in the middle of the table. On beginning his turn, each player may first take one card from the pile. Then that player may ask for one card from the player to his right (as in "Go Fish"), and the player on the right, if he has it, must give it to the player who asked. Then the player whose turn it is may discard the card 10 or any two cards in his hand that add up to 10. The player who first discards all the cards from his hand is the winner.

Variations:
* Deal more than seven cards.
* Deal fewer cards if there are a lot of players or the players are very young.
* Allow players to discard *three* cards that add up to 10.
* Instead of ten, players discard cards that add up to 9, 8, 11, or some other number.
  Use the face cards Jack, Queen, and King for 11, 12, and 13 respectively.

Some Went Hiding

You need: The same number of small objects as the sum you are studying. For example, to study the sums with 5, you need 5 objects (marbles, blocks, or whatever).

Rules: The first player shows the objects but quickly hides some of them behind his back without showing how many. Then he shows the remaining objects to the next player, who has to say how many "went hiding." If the player gives the right answer, it is then his turn to hide some and ask the next player to answer. If he gives a wrong answer, he forfeits his turn. This game appeals best to young children.

Variation: Instead of getting a turn to hide objects, the player who answers correctly may gain points or other rewards for the right answer.

Addition (or Subtraction) Challenge

You need: A standard deck of playing cards from which you remove the face cards and perhaps also some of the other higher-numbered cards, such as tens, nines, and eights. Alternatively, a set of dominoes works well for children who do not yet know their numbers beyond 12.

Rules: In each round, each player is dealt two cards face up, and has to calculate the sum or difference (add/subtract). The player with the highest sum or difference gets all the cards from the other players. After enough rounds have been played to use all of the cards, the player with the most cards wins.

If two or more players have the same sum, then those players get an additional two cards and use those to resolve the tie.

Variations:
* This game is easily adapted for subtraction, multiplication, and fractions.
* You can also use dominoes instead of two playing cards.

Any ***board game*** where you move the piece by rolling two dice also works to practice addition.

Helpful Resources on the Internet

You can also access this list of links at https://links.mathmammoth.com/gr1ch1

Video Lessons - Addition Within 0-10
A set of videos by Maria that match the topics in this chapter.
https://www.mathmammoth.com/videos/grade_1/1st-grade-math-videos.php#addition

Basic Addition & Subtraction Facts — online practice
An ad-free online program to practice basic addition and subtraction at MathMammoth.com website.
https://www.mathmammoth.com/practice/addition-single-digit.php

Hidden Picture Addition Game
Click on the correct answer to each addition problem and uncover a hidden picture.
https://www.mathmammoth.com/practice/mystery-picture#min=0&max=6

Kid's Addition Quiz
A set of five problems. Choose the maximum for the sum from the list of numbers below the quiz.
http://www.thegreatmartinicompany.com/Math-Quick-Quiz/addition-kid-quiz.html

Kids Compare Numbers from Mr. Martini's Classroom
Compare two numbers. Press the number below to choose the biggest number that will appear.
http://www.thegreatmartinicompany.com/Kids-Math/compare-number.html

Balloon Pop Math – Compare Numbers
Click on "is greater than", "is equal to", or "is less than" to compare the two given numbers.
https://www.sheppardsoftware.com/mathgames/earlymath/BPGreatLessEqualWords.htm

Line Jumper
Solve addition problems on a number line.
https://www.funbrain.com/games/line-jumper

Math Lines
Practice adding in this fun game. First, choose the number to practice. Then, shoot the numbered marble from the cannon into a numbered marble such that the numbers total the target number.
https://www.mathnook.com/math/math-lines-6.html

Number Twins
First, click on the number that you want to practice. Then, match pairs of balls that add up to that number.
https://www.sheppardsoftware.com/mathgames/numbertwins/numbertwins_add_10.htm

Balloon Flight Addition
Practice mental addition of three single-digit numbers with this interactive online game.
https://www.ictgames.com/mobilePage/balloon/index.html

Save the Whale
Find how much the given "pipe" length needs added to make it 10 long and save the whale.
https://www.ictgames.com/saveTheWhale/index.html

Addition with Missing Numbers
Match the missing number to its correct value.
https://www.sheppardsoftware.com/mathgames/matching/AdditionX.htm

Balloon Pop - Add
Pop the balloons in order: from the smallest sum to the largest sum.
https://www.sheppardsoftware.com/mathgames/numberballoons/NumberBalloons_add_level1.htm

Fill in the block on each line to complete the pattern.

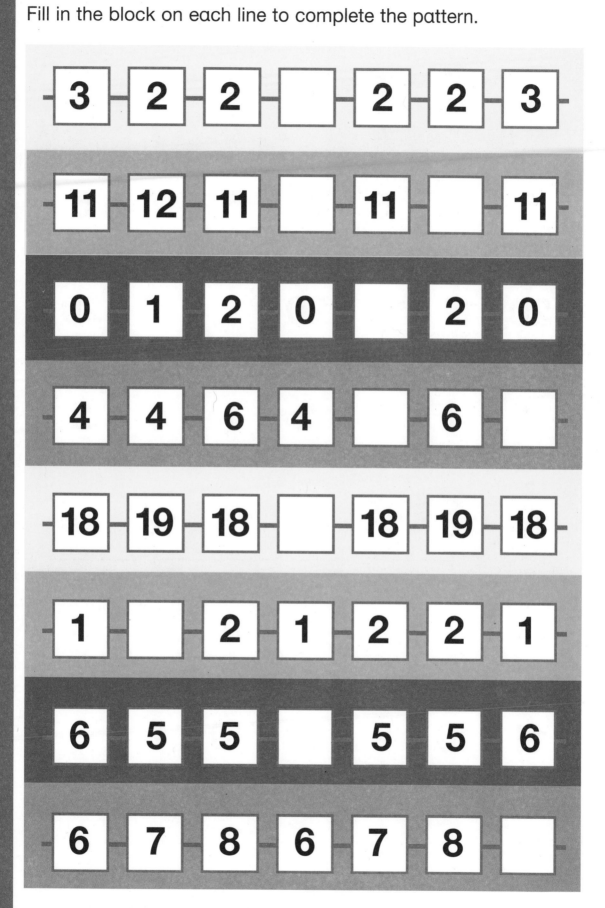

Fill in the block on each line to complete the pattern.

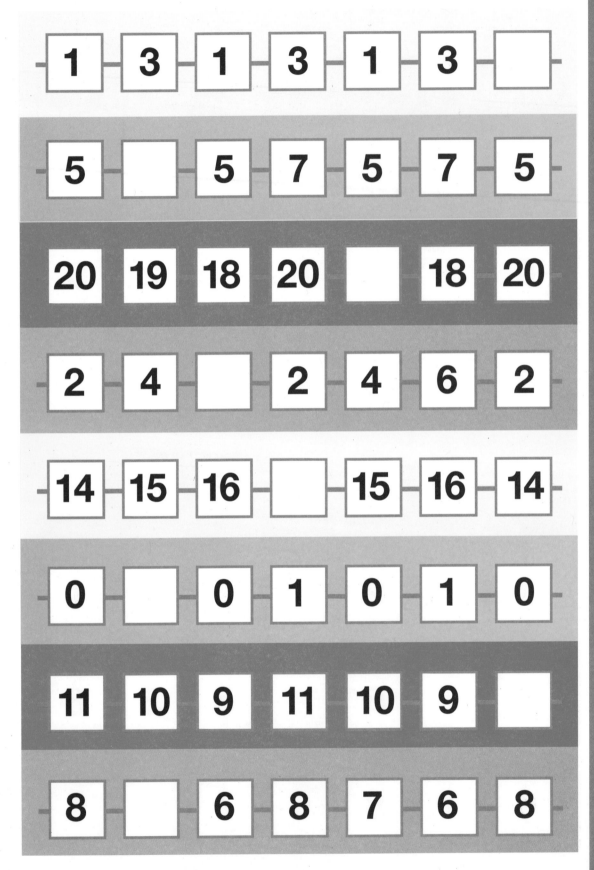

1 3 1 3 1 3 [ ]

5 [ ] 5 7 5 7 5

20 19 18 20 [ ] 18 20

2 4 [ ] 2 4 6 2

14 15 16 [ ] 15 16 14

0 [ ] 0 1 0 1 0

11 10 9 11 10 9 [ ]

8 [ ] 6 8 7 6 8

Two Groups and a Total

1. Make two groups.

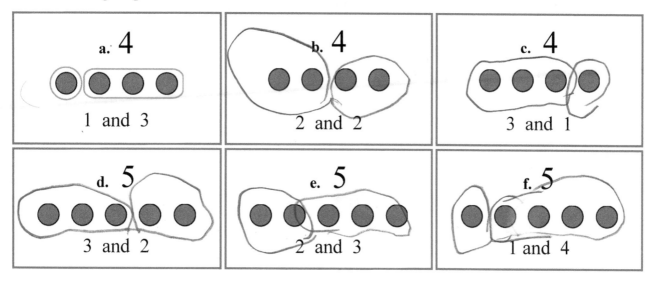

a. 4

1 and 3

b. 4

2 and 2

c. 4

3 and 1

d. 5

3 and 2

e. 5

2 and 3

f. 5

1 and 4

2. Make two groups. Write how many are in the second group.

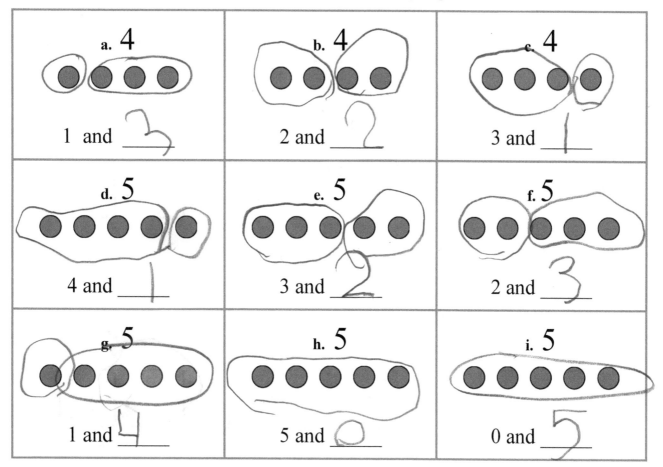

a. 4

1 and 3

b. 4

2 and 2

c. 4

3 and 1

d. 5

4 and 1

e. 5

3 and 2

f. 5

2 and 3

g. 5

1 and 4

h. 5

5 and 0

i. 5

0 and 5

3. Draw as many dots as the number shows. Then divide them into two groups.
   (There are many ways to do this.) Write how many are in each group.

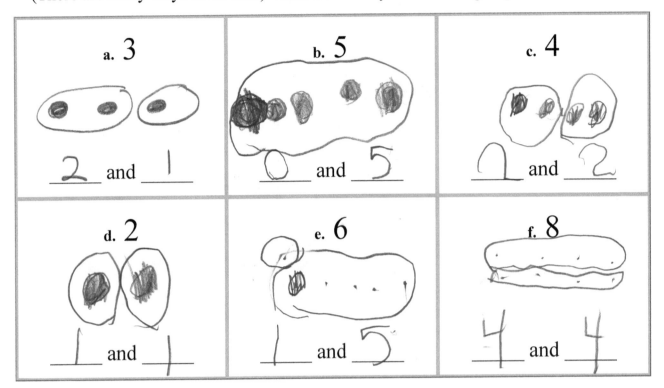

a. 3   _2_ and _1_

b. 5   ___ and _5_

c. 4   _1_ and _2_

d. 2   ___ and _1_

e. 6   ___ and _5_

f. 8   _4_ and _4_

4. The number at the top is the total. Draw the missing dots on the face of the blank dice.
   Write on the lines how many dots are on the face of each dice.

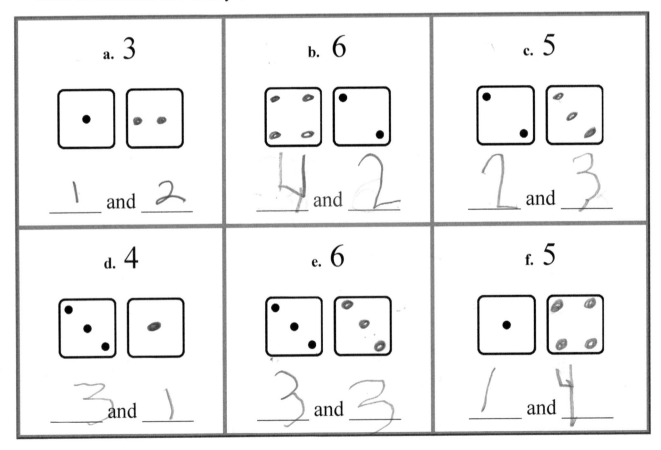

a. 3   _1_ and _2_

b. 6   _4_ and _2_

c. 5   _2_ and _3_

d. 4   _3_ and _1_

e. 6   _3_ and _3_

f. 5   _1_ and _4_

2 and 2   [4]

"Two and two makes four."

1 and 4   [5]

"One and four makes five."

5. Write how many are in each group. Write the total in the box.

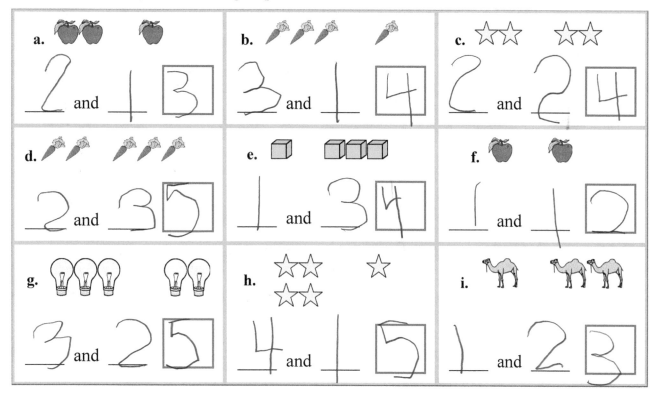

a. 2 and 3 [3]

b. 3 and 1 [4]

c. 2 and 2 [4]

d. 2 and 3 [5]

e. 1 and 3 [4]

f. 1 and 1 [2]

g. 3 and 2 [5]

h. 4 and 1 [5]

i. 1 and 2 [3]

6. Draw circles for each number. Write the total in the box.

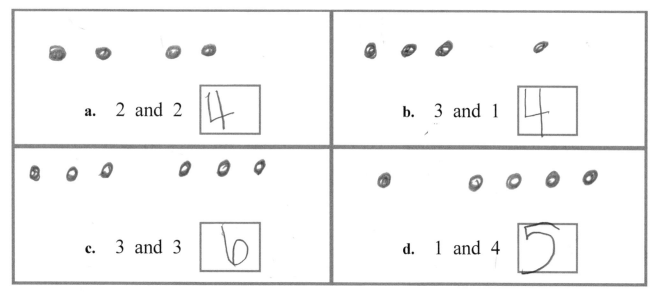

a. 2 and 2 [4]

b. 3 and 1 [4]

c. 3 and 3 [6]

d. 1 and 4 [5]

21

Learn the Symbols + and =

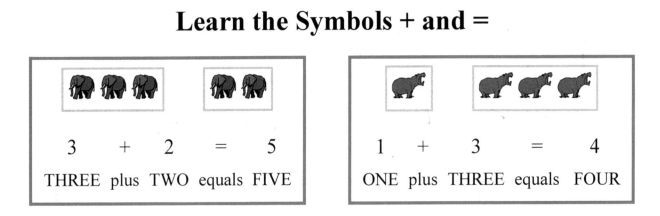

3 + 2 = 5

THREE plus TWO equals FIVE

1 + 3 = 4

ONE plus THREE equals FOUR

1. Fill in the numbers. Add. Read the additions aloud using "plus" and "equals".

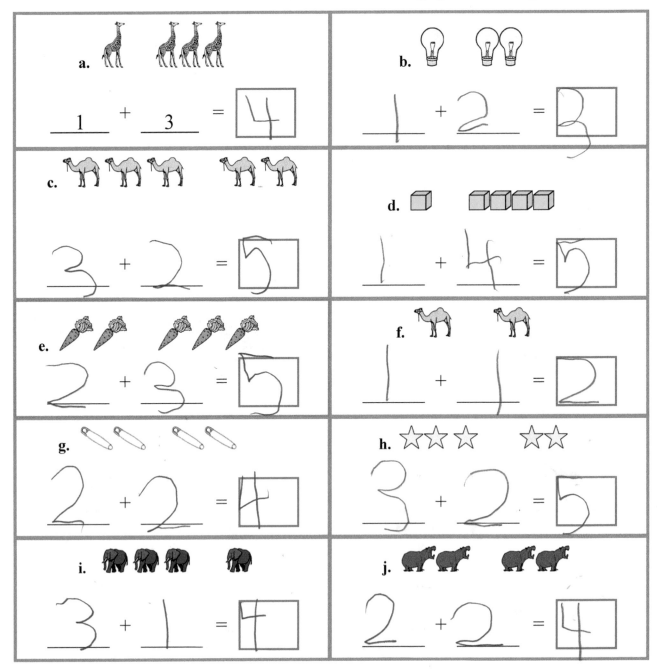

a. _1_ + _3_ = 4

b. 1 + 2 = 3

c. 3 + 2 = 5

d. 1 + 4 = 5

e. 2 + 3 = 5

f. 1 + 1 = 2

g. 2 + 2 = 4

h. 3 + 2 = 5

i. 3 + 1 = 4

j. 2 + 2 = 4

2. Write the numbers. Add. Read the additions aloud using "plus" and "equals".

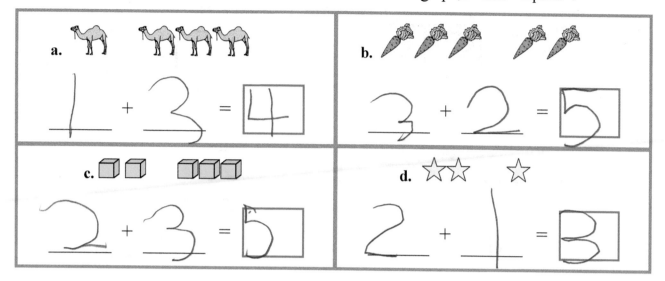

a. $1 + 3 = 4$

b. $3 + 2 = 5$

c. $2 + 3 = 5$

d. $2 + 1 = 3$

3. Add with zero.

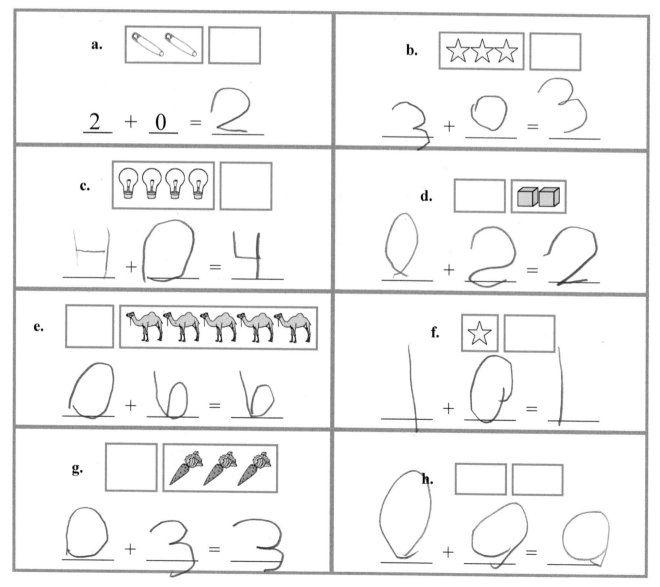

a. $2 + 0 = 2$

b. $3 + 0 = 3$

c. $4 + 0 = 4$

d. $0 + 2 = 2$

e. $0 + 6 = 6$

f. $1 + 0 = 1$

g. $0 + 3 = 3$

h. $0 + 0 = 0$

23

4. Write how many dots. Then add.

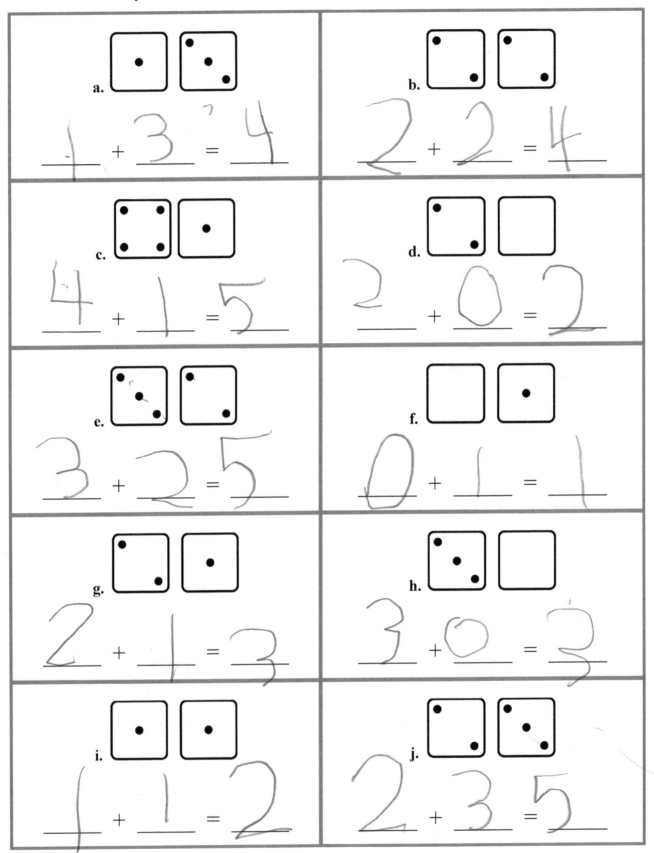

a. $\underline{1} + \underline{3} = \underline{4}$

b. $\underline{2} + \underline{2} = \underline{4}$

c. $\underline{4} + \underline{1} = \underline{5}$

d. $\underline{2} + \underline{0} = \underline{2}$

e. $\underline{3} + \underline{2} = \underline{5}$

f. $\underline{0} + \underline{1} = \underline{1}$

g. $\underline{2} + \underline{1} = \underline{3}$

h. $\underline{3} + \underline{0} = \underline{3}$

i. $\underline{1} + \underline{1} = \underline{2}$

j. $\underline{2} + \underline{3} = \underline{5}$

Addition Practice 1

1. In the second box, draw enough things to show the second number. Then add.

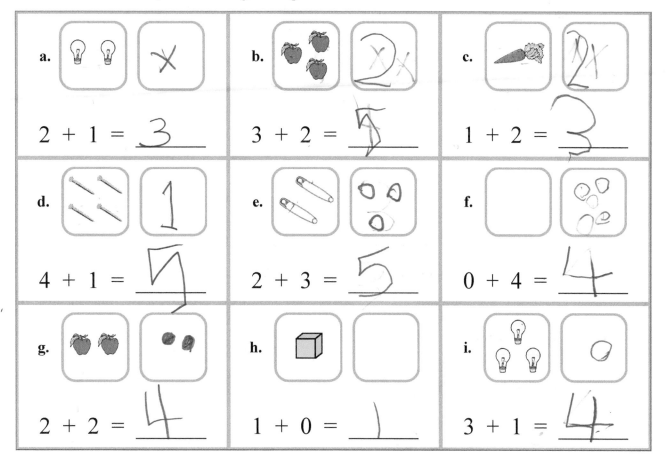

a.   2 + 1 = 3

b.   3 + 2 = 5

c.   1 + 2 = 3

d.   4 + 1 = 5

e.   2 + 3 = 5

f.   0 + 4 = 4

g.   2 + 2 = 4

h.   1 + 0 = 1

i.   3 + 1 = 4

2. Draw dots in each box for the numbers. Then add.

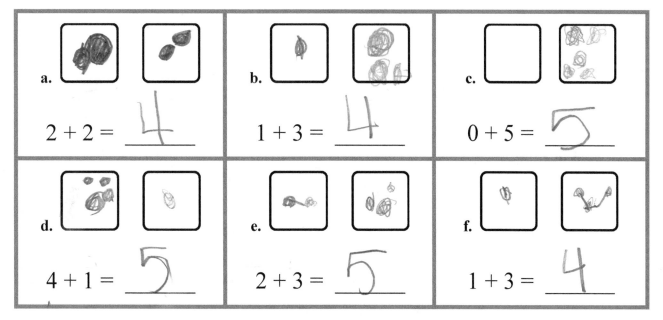

a.   2 + 2 = 4

b.   1 + 3 = 4

c.   0 + 5 = 5

d.   4 + 1 = 5

e.   2 + 3 = 5

f.   1 + 3 = 4

3. Add. If you want to, you can draw circles or sticks to help you.

a. 1 + 2 = 3	b. 3 + 0 = 3	c. 2 + 2 = 4
d. 2 + 3 = 5	e. 1 + 4 = 5	f. 0 + 5 = 5
g. 3 + 2 = 5	h. 2 + 1 = 3	i. 4 + 1 = 5

4. Add in both orders! Notice: the answer is the same. You can draw marbles to help.

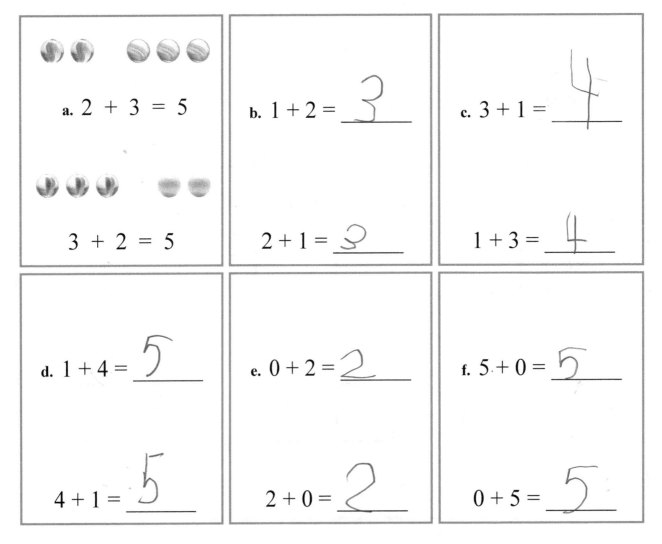

a. 2 + 3 = 5 3 + 2 = 5	b. 1 + 2 = 3 2 + 1 = 3	c. 3 + 1 = 4 1 + 3 = 4
d. 1 + 4 = 5 4 + 1 = 5	e. 0 + 2 = 2 2 + 0 = 2	f. 5 + 0 = 5 0 + 5 = 5

Which Is More?

The symbols < and > are like a "hungry alligator's mouth."
The mouth always opens towards the **bigger** number.

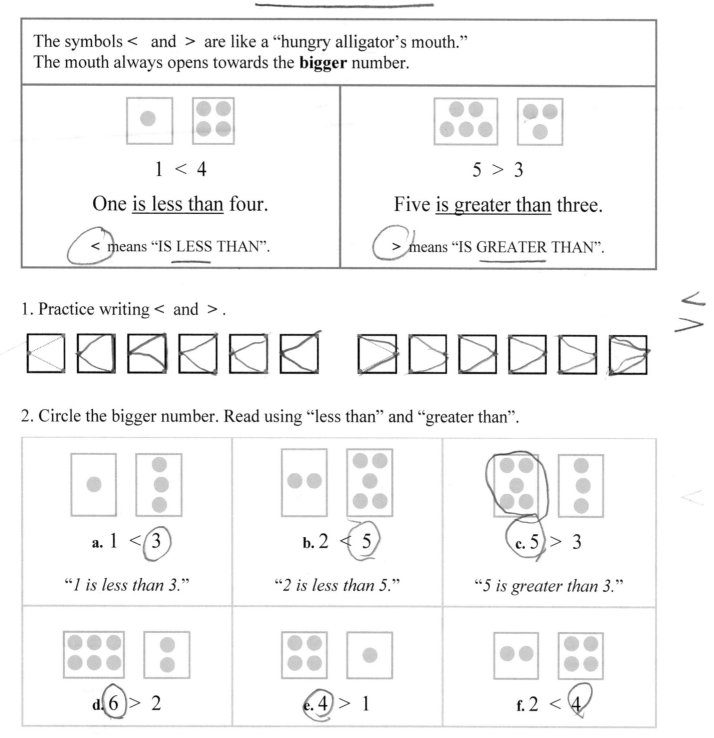

1 < 4

One <u>is less than</u> four.

< means "IS LESS THAN".

5 > 3

Five <u>is greater than</u> three.

> means "IS GREATER THAN".

1. Practice writing < and > .

2. Circle the bigger number. Read using "less than" and "greater than".

a. 1 < ③

"1 is less than 3."

b. 2 < ⑤

"2 is less than 5."

c. ⑤ > 3

"5 is greater than 3."

d. ⑥ > 2

e. ④ > 1

f. 2 < ④

3. Circle the bigger number. Read using "less than" and "greater than".

| **a.** ⑥ > 0 | **b.** 3 < ④ | **c.** 4 < ⑤ | **d.** ④ > 3 |
| **e.** 1 < ② | **f.** ② > 1 | **g.** 3 < ⑤ | **h.** 0 < ④ |

4. Write < or > in the box.

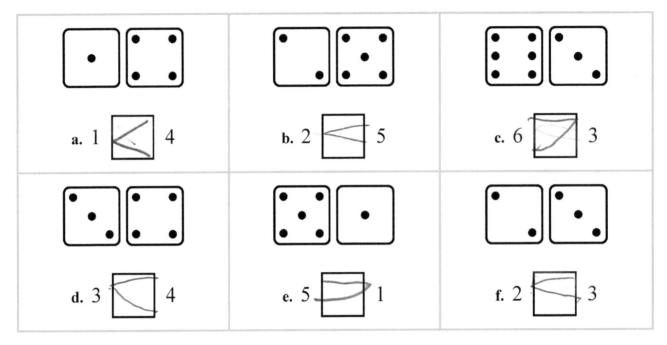

a. 1 < 4

b. 2 < 5

c. 6 > 3

d. 3 < 4

e. 5 > 1

f. 2 < 3

5. Write < or > between the numbers. You can draw circles to help you.

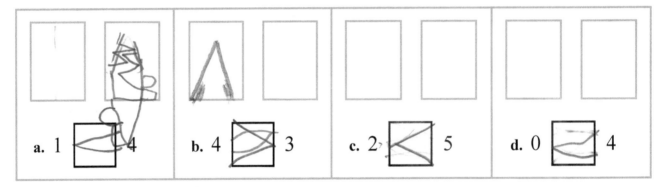

a. 1 < 4

b. 4 > 3

c. 2 < 5

d. 0 < 4

6. Write < or > between the two numbers.

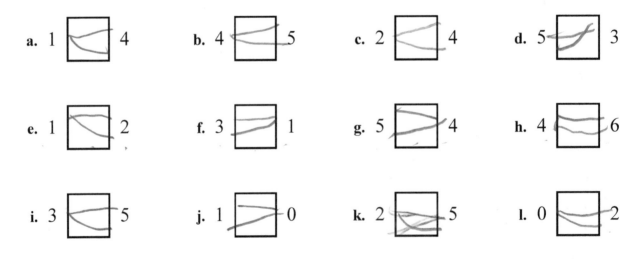

a. 1 < 4

b. 4 < 5

c. 2 < 4

d. 5 > 3

e. 1 < 2

f. 3 > 1

g. 5 > 4

h. 4 < 6

i. 3 < 5

j. 1 > 0

k. 2 < 5

l. 0 < 2

Missing Items

Something is missing from the addition.
The TOTAL is not missing. The total is 5.

How many are in the second group? That is what is missing!

There should be a total of 5 dots. Draw 4 in the face of the second dice.

5

[dice] + [dice]

1 + _____

There should be a total of 4 dots. The face of the second dice has two. There are none on the face of the first dice, so you need to draw them.

Read: "2 plus what number makes 4?"
   or, "2 and how many more makes 4?"
   or, "What number and 2 makes 4?"

4

[dice] + [dice]

_____ + 2

1. Complete the addition. Draw the missing dots. The total is on top.

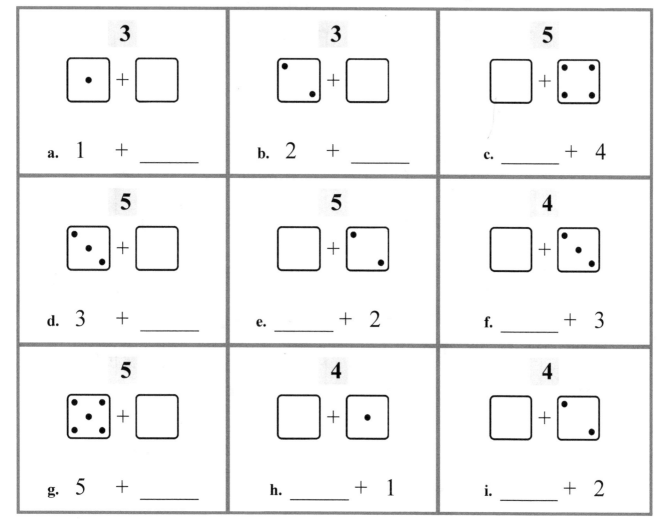

3

a. 1 + _____

3

b. 2 + _____

5

c. _____ + 4

5

d. 3 + _____

5

e. _____ + 2

4

f. _____ + 3

5

g. 5 + _____

4

h. _____ + 1

4

i. _____ + 2

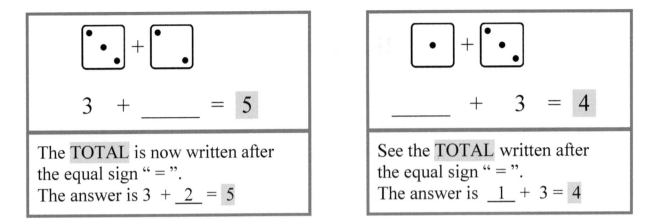

The TOTAL is now written after the equal sign " = ".
The answer is 3 + __2__ = 5

See the TOTAL written after the equal sign " = ".
The answer is __1__ + 3 = 4

2. Draw more dots to show the missing number. Write the missing number.

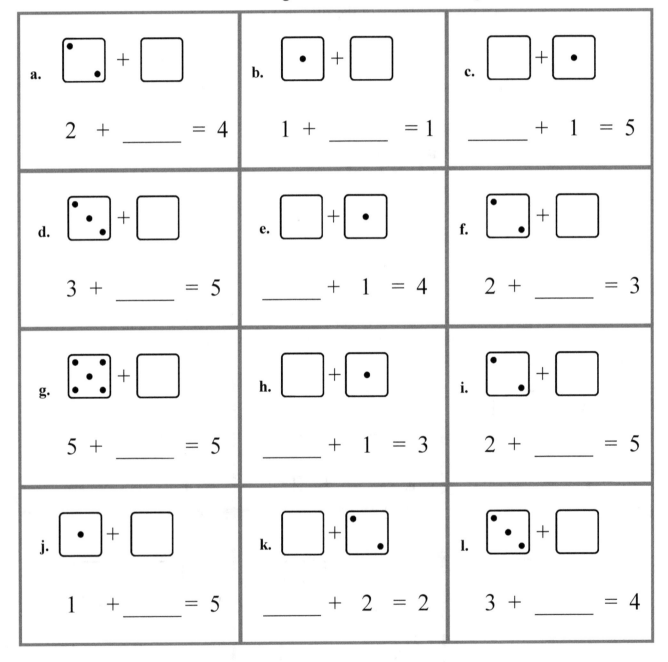

3. Draw dots in the empty box for the missing number. Read the problems aloud:
   "2 plus how many makes 4?"

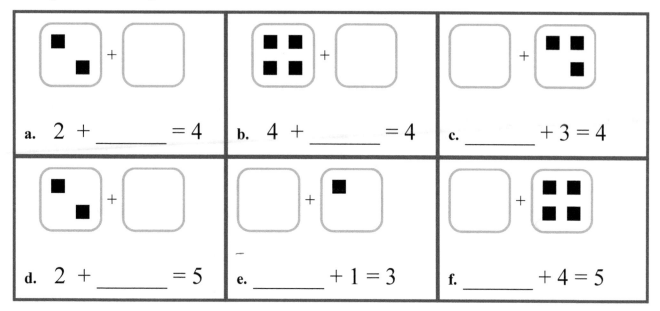

a. $2 + \underline{\hspace{1.5cm}} = 4$ b. $4 + \underline{\hspace{1.5cm}} = 4$ c. $\underline{\hspace{1.5cm}} + 3 = 4$

d. $2 + \underline{\hspace{1.5cm}} = 5$ e. $\underline{\hspace{1.5cm}} + 1 = 3$ f. $\underline{\hspace{1.5cm}} + 4 = 5$

There are no dots on the face of either dice.

The face of the first dice is missing its dots. The face of the second is *supposed* to have none, since there is a zero below it.

Draw 4 dots on the face of the first dice, because $\underline{4} + 0 = \boxed{4}$.

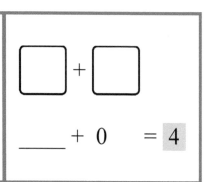

$\underline{\hspace{1.5cm}} + 0 \quad = \boxed{4}$

4. Draw dots in the boxes for the missing numbers. Notice that some boxes are supposed to have zero dots.

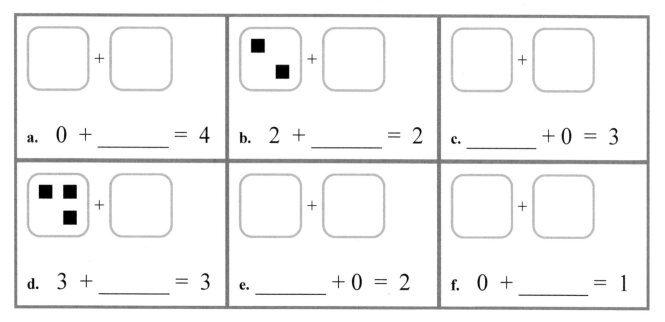

a. $0 + \underline{\hspace{1.5cm}} = 4$ b. $2 + \underline{\hspace{1.5cm}} = 2$ c. $\underline{\hspace{1.5cm}} + 0 = 3$

d. $3 + \underline{\hspace{1.5cm}} = 3$ e. $\underline{\hspace{1.5cm}} + 0 = 2$ f. $0 + \underline{\hspace{1.5cm}} = 1$

5. Draw dots to illustrate each addition problem. Find what number is missing.

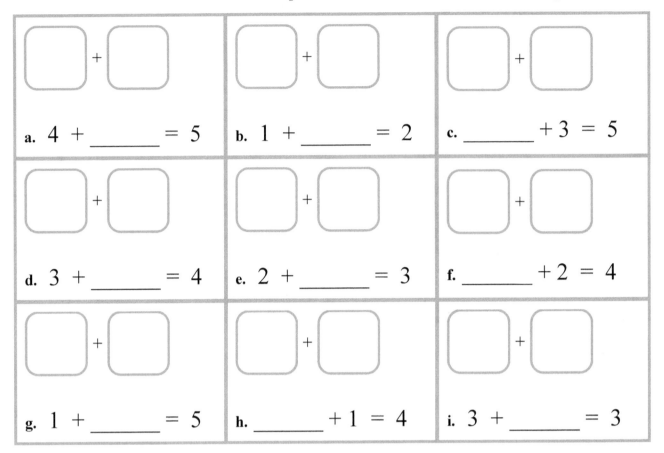

a. 4 + _____ = 5

b. 1 + _____ = 2

c. _____ + 3 = 5

d. 3 + _____ = 4

e. 2 + _____ = 3

f. _____ + 2 = 4

g. 1 + _____ = 5

h. _____ + 1 = 4

i. 3 + _____ = 3

6. Solve. Now, the missing number goes inside the shape. You can draw dots to help you. Remember, the number after the " = " sign is the total.

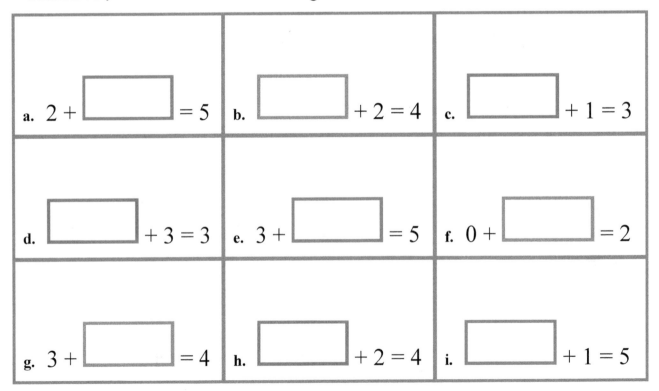

a. 2 + ☐ = 5

b. ☐ + 2 = 4

c. ☐ + 1 = 3

d. ☐ + 3 = 3

e. 3 + ☐ = 5

f. 0 + ☐ = 2

g. 3 + ☐ = 4

h. ☐ + 2 = 4

i. ☐ + 1 = 5

7. Practice "normal" addition.

a. 1 + 1 = _____

    2 + 1 = _____

b. 4 + 0 = _____

    3 + 1 = _____

c. 1 + 4 = _____

    2 + 2 = _____

d. 2 + 3 = _____

    1 + 4 = _____

e. 0 + 5 = _____

    1 + 2 = _____

f. 3 + 2 = _____

    4 + 1 = _____

8. Find the missing number. The marbles illustrate the total. Notice the patterns!

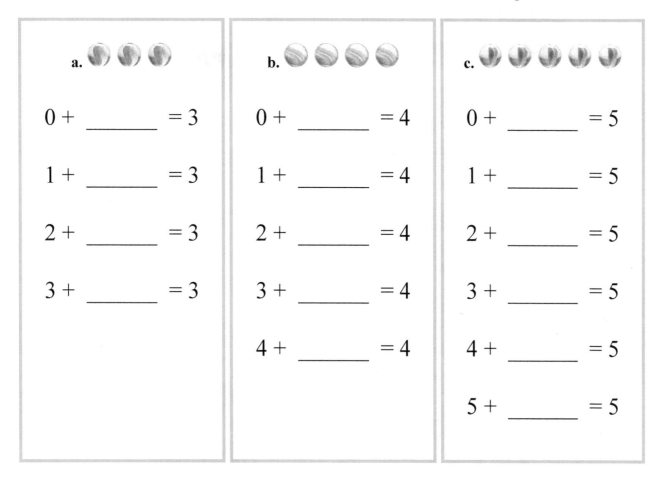

a.

0 + _____ = 3

1 + _____ = 3

2 + _____ = 3

3 + _____ = 3

b.

0 + _____ = 4

1 + _____ = 4

2 + _____ = 4

3 + _____ = 4

4 + _____ = 4

c.

0 + _____ = 5

1 + _____ = 5

2 + _____ = 5

3 + _____ = 5

4 + _____ = 5

5 + _____ = 5

Sums with 5

1. Here are some different ways to group five elephants into two groups. The " | " symbol separates the two groups. Write the addition sentences.

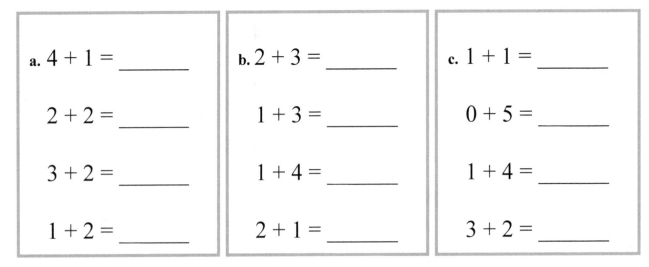

_____ + _____ = _____   _____ + _____ = _____

_____ + _____ = _____   _____ + _____ = _____

_____ + _____ = _____   _____ + _____ = _____

2. Add.

a. 4 + 1 = _____   b. 2 + 3 = _____   c. 1 + 1 = _____

2 + 2 = _____   1 + 3 = _____   0 + 5 = _____

3 + 2 = _____   1 + 4 = _____   1 + 4 = _____

1 + 2 = _____   2 + 1 = _____   3 + 2 = _____

3. Play "5 Out" *and/or* "Some Went Hiding" with 5 objects (see the introduction).

4. **Drill.** Don't write the answers in the boxes, but just solve them in your head.

$1 + \square = 5$ $\qquad$ $4 + \square = 5$ $\qquad$ $\square + 2 = 5$ $\qquad$ $\square + 3 = 5$

$2 + \square = 5$ $\qquad$ $3 + \square = 5$ $\qquad$ $\square + 0 = 5$ $\qquad$ $\square + 1 = 5$

$0 + \square = 5$ $\qquad$ $5 + \square = 5$ $\qquad$ $\square + 4 = 5$ $\qquad$ $\square + 5 = 5$

5. Add. Compare the problems in each group. You can draw more shapes to help you with the additions.

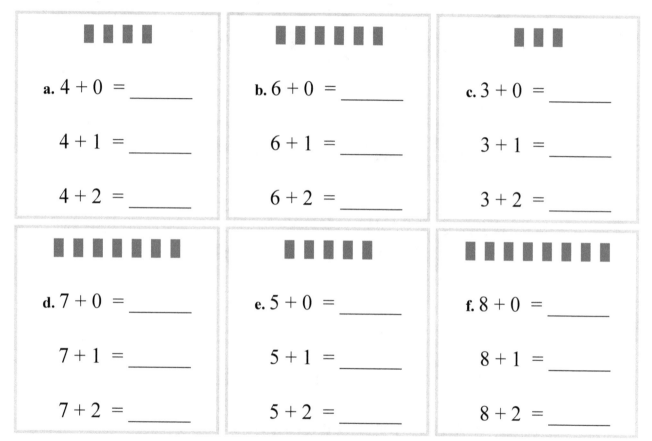

a. $4 + 0 =$ _____

$4 + 1 =$ _____

$4 + 2 =$ _____

b. $6 + 0 =$ _____

$6 + 1 =$ _____

$6 + 2 =$ _____

c. $3 + 0 =$ _____

$3 + 1 =$ _____

$3 + 2 =$ _____

d. $7 + 0 =$ _____

$7 + 1 =$ _____

$7 + 2 =$ _____

e. $5 + 0 =$ _____

$5 + 1 =$ _____

$5 + 2 =$ _____

f. $8 + 0 =$ _____

$8 + 1 =$ _____

$8 + 2 =$ _____

6. Draw more things to illustrate the missing number. Complete the addition sentence.

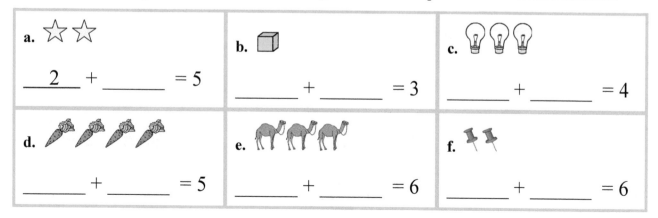

a. ____2____ + _____ = 5

b. _____ + _____ = 3

c. _____ + _____ = 4

d. _____ + _____ = 5

e. _____ + _____ = 6

f. _____ + _____ = 6

Sums with 6

1. Here are some different ways to group six hippos into two groups.
   Write the addition sentences.

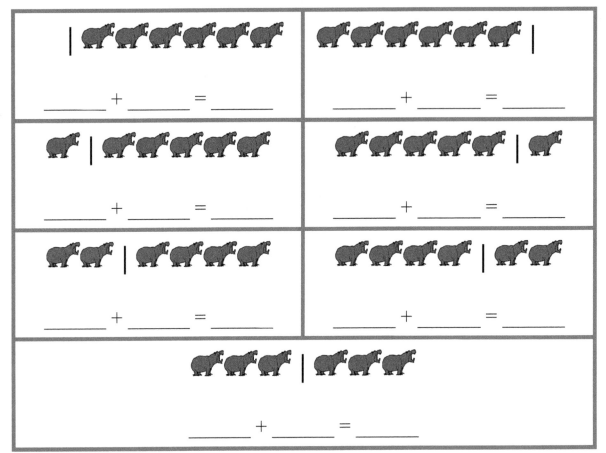

2. Play "6 Out" *and/or* "Some Went Hiding" with 6 objects (see the introduction).

3. **Drill.** Don't write the answers but just solve them in your head.

$1 + \square = 6$ $4 + \square = 6$ $\square + 2 = 6$ $\square + 3 = 6$

$2 + \square = 6$ $3 + \square = 6$ $\square + 0 = 6$ $\square + 1 = 6$

$6 + \square = 6$ $5 + \square = 6$ $\square + 4 = 6$ $\square + 5 = 6$

4. Add.

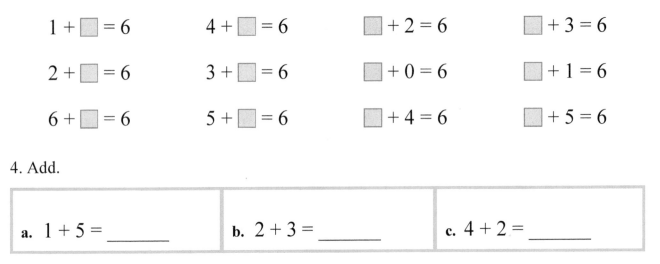

| a. $1 + 5 = $ _____ | b. $2 + 3 = $ _____ | c. $4 + 2 = $ _____ |

5. Draw more little boxes to illustrate the missing number.

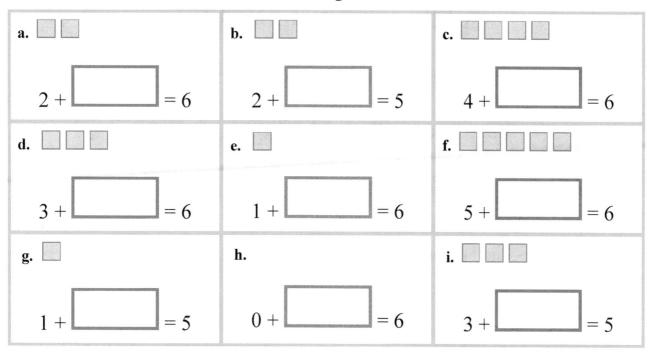

a. $2 + \boxed{} = 6$

b. $2 + \boxed{} = 5$

c. $4 + \boxed{} = 6$

d. $3 + \boxed{} = 6$

e. $1 + \boxed{} = 6$

f. $5 + \boxed{} = 6$

g. $1 + \boxed{} = 5$

h. $0 + \boxed{} = 6$

i. $3 + \boxed{} = 5$

6. Jack and Jill share 5 cucumbers and 6 lemons in different ways. Find how many Jill gets.
   You can cover the cucumbers or lemons with your hand to help.

a. 5

Jack gets:	Left for Jill:
2	
1	
5	
3	
0	
4	

b. 6

Jack gets:	Left for Jill:
1	
4	
5	
0	
2	
3	

7. Add.

$2 + 3 =$ _____

$4 + 1 =$ _____

$3 + 3 =$ _____

$4 + 2 =$ _____

$1 + 3 =$ _____

$1 + 5 =$ _____

$2 + 2 =$ _____

$2 + 4 =$ _____

Adding on a Number Line

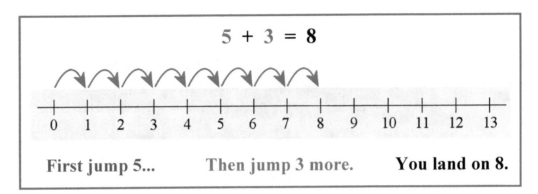

5 + 3 = 8

First jump 5...   Then jump 3 more.   **You land on 8.**

1. Draw the jumps to illustrate the addition and find the answer.
   You can use a different color for each number when you draw the jumps.

a. $5 + 2 =$ _____

b. $4 + 1 =$ _____

c. $6 + 3 =$ _____

d. $9 + 1 =$ _____

e. $7 + 3 =$ _____

f. $4 + 3 =$ _____

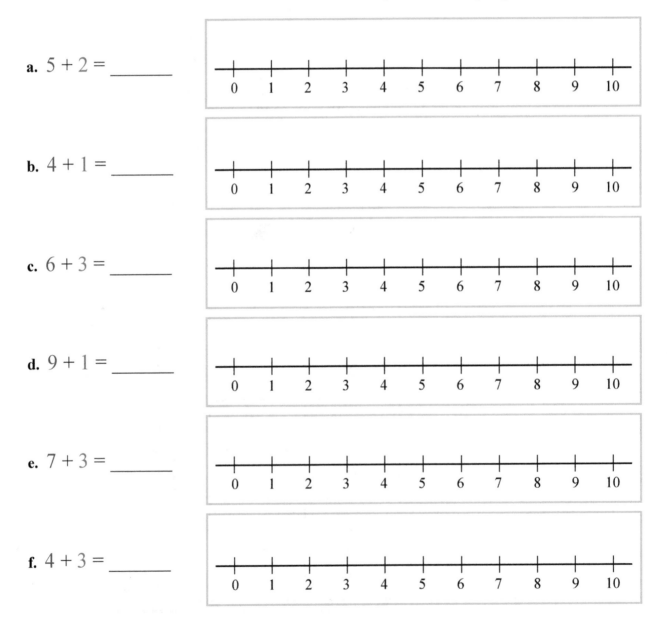

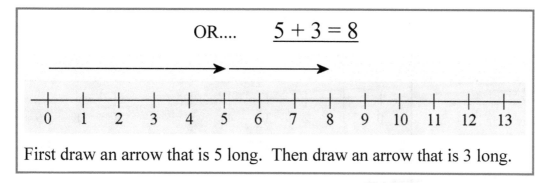

OR....   5 + 3 = 8

First draw an arrow that is 5 long. Then draw an arrow that is 3 long.

2. Write the addition sentence.

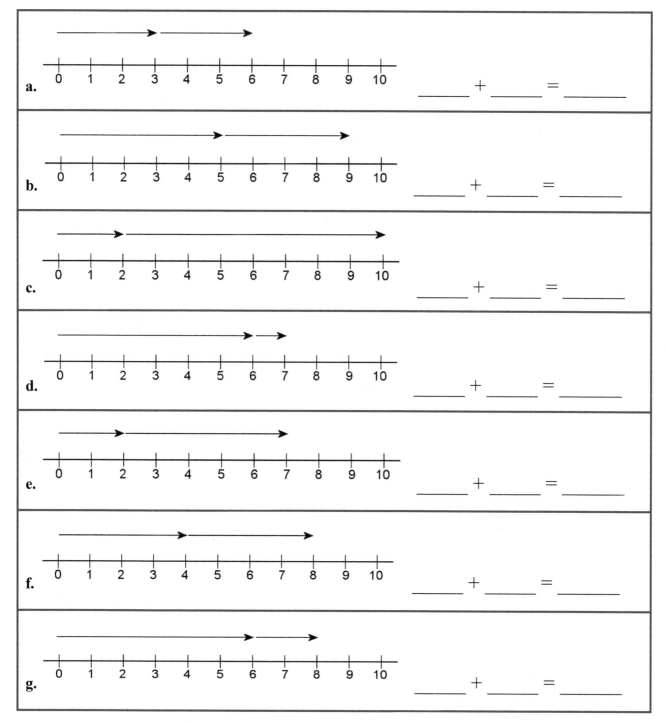

a. _____ + _____ = _____

b. _____ + _____ = _____

c. _____ + _____ = _____

d. _____ + _____ = _____

e. _____ + _____ = _____

f. _____ + _____ = _____

g. _____ + _____ = _____

3. Draw arrows (or jumps) to show the addition.

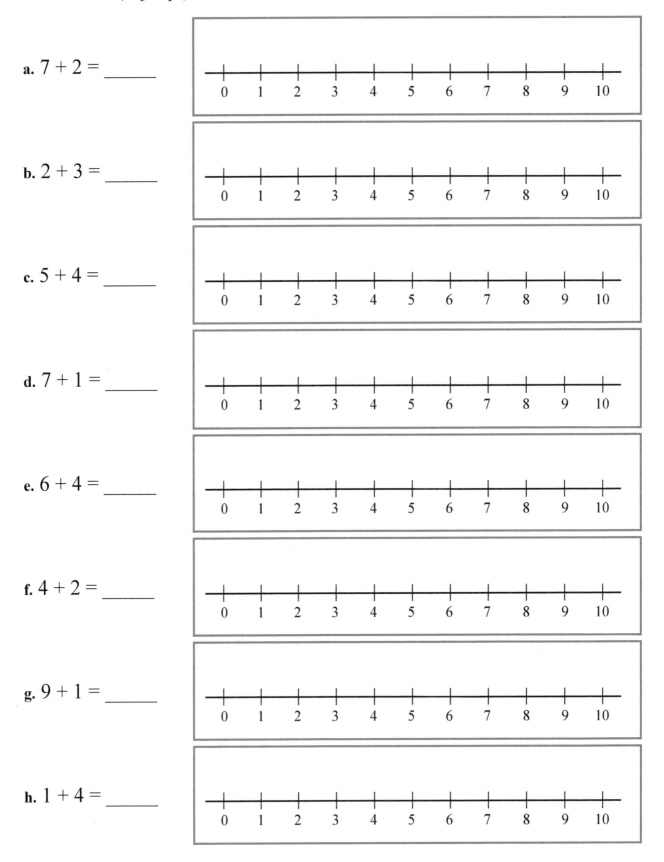

a. 7 + 2 = _____

        0   1   2   3   4   5   6   7   8   9   10

b. 2 + 3 = _____

        0   1   2   3   4   5   6   7   8   9   10

c. 5 + 4 = _____

        0   1   2   3   4   5   6   7   8   9   10

d. 7 + 1 = _____

        0   1   2   3   4   5   6   7   8   9   10

e. 6 + 4 = _____

        0   1   2   3   4   5   6   7   8   9   10

f. 4 + 2 = _____

        0   1   2   3   4   5   6   7   8   9   10

g. 9 + 1 = _____

        0   1   2   3   4   5   6   7   8   9   10

h. 1 + 4 = _____

        0   1   2   3   4   5   6   7   8   9   10

4. Write the addition sentence for each picture.
   If the child is not familiar with numbers greater than 10, you can skip these.

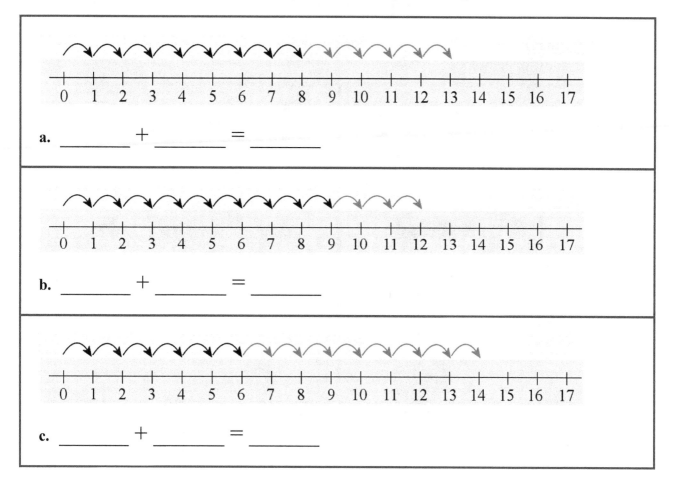

a. _____ + _____ = _____

b. _____ + _____ = _____

c. _____ + _____ = _____

5. Add "1", add "2" to the number. Use the number line to help.

a.	b.	c.	d.
7 + 1 = _____	5 + 1 = _____	6 + 1 = _____	8 + 1 = _____
7 + 2 = _____	5 + 2 = _____	6 + 2 = _____	8 + 2 = _____

e.	f.	g.	h.
10 + 1 = _____	12 + 1 = _____	13 + 1 = _____	11 + 1 = _____
10 + 2 = _____	12 + 2 = _____	13 + 2 = _____	11 + 2 = _____

Sums with 7

1. Here are some different ways to group seven marbles into two groups.
   Write the addition sentences.

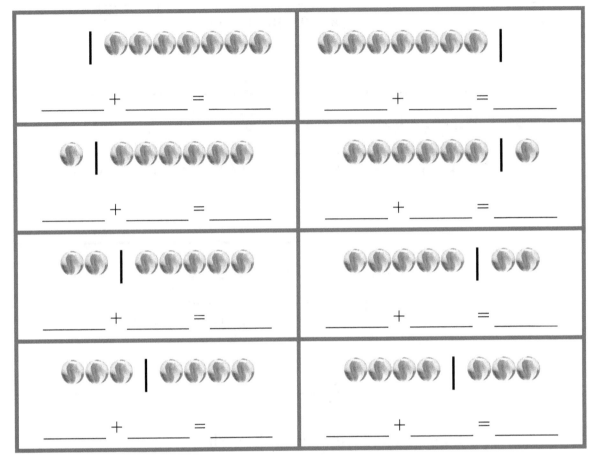

_____ + _____ = _____	_____ + _____ = _____
_____ + _____ = _____	_____ + _____ = _____
_____ + _____ = _____	_____ + _____ = _____
_____ + _____ = _____	_____ + _____ = _____

2. **Drill.** Don't write the answers here. Just solve them in your head.

$5 + \square = 7$ $2 + \square = 7$ $6 + \square = 7$ $\square + 3 = 7$ $\square + 7 = 7$

$3 + \square = 7$ $1 + \square = 7$ $0 + \square = 7$ $\square + 2 = 7$ $\square + 1 = 7$

$7 + \square = 7$ $4 + \square = 7$ $4 + \square = 7$ $\square + 6 = 7$ $\square + 5 = 7$

3. Add.

a.	b.	c.	d.
$3 + 3 =$ _____	$5 + 2 =$ _____	$6 + 1 =$ _____	$2 + 5 =$ _____
$3 + 4 =$ _____	$4 + 2 =$ _____	$4 + 3 =$ _____	$4 + 2 =$ _____

4. Play "7 Out" *and/or* "Some Went Hiding" with 7 objects (see the introduction).

5. Fill in the missing numbers. You may draw dots to help. Notice the patterns!

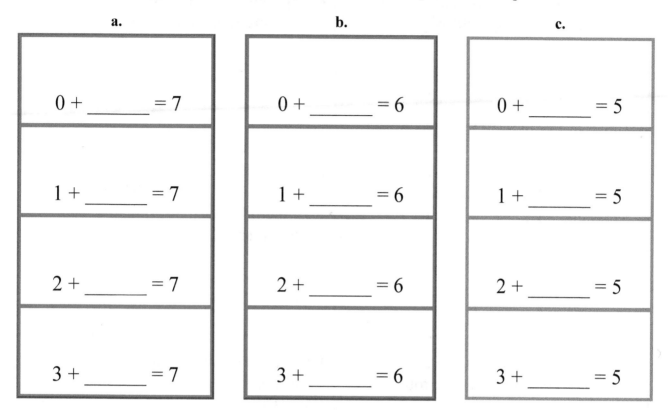

a.		b.		c.
$0 + \rule{1cm}{0.4pt} = 7$		$0 + \rule{1cm}{0.4pt} = 6$		$0 + \rule{1cm}{0.4pt} = 5$
$1 + \rule{1cm}{0.4pt} = 7$		$1 + \rule{1cm}{0.4pt} = 6$		$1 + \rule{1cm}{0.4pt} = 5$
$2 + \rule{1cm}{0.4pt} = 7$		$2 + \rule{1cm}{0.4pt} = 6$		$2 + \rule{1cm}{0.4pt} = 5$
$3 + \rule{1cm}{0.4pt} = 7$		$3 + \rule{1cm}{0.4pt} = 6$		$3 + \rule{1cm}{0.4pt} = 5$

6. This is a new way to write addition! The answer goes *under* the line.

a.
$$\begin{array}{r} 4 \\ + 3 \\ \hline 7 \end{array}$$

b.
$$\begin{array}{r} 1 \\ + 5 \\ \hline \end{array}$$

c.
$$\begin{array}{r} 5 \\ + 0 \\ \hline \end{array}$$

d.
$$\begin{array}{r} 4 \\ + 1 \\ \hline \end{array}$$

e.
$$\begin{array}{r} 4 \\ + 0 \\ \hline \end{array}$$

f.
$$\begin{array}{r} 2 \\ + 5 \\ \hline \end{array}$$

g.
$$\begin{array}{r} 0 \\ + 3 \\ \hline \end{array}$$

h.
$$\begin{array}{r} 1 \\ + 3 \\ \hline \end{array}$$

i.
$$\begin{array}{r} 3 \\ + 3 \\ \hline \end{array}$$

j.
$$\begin{array}{r} 2 \\ + 2 \\ \hline \end{array}$$

k.
$$\begin{array}{r} 4 \\ + 2 \\ \hline \end{array}$$

l.
$$\begin{array}{r} 2 \\ + 0 \\ \hline \end{array}$$

m.
$$\begin{array}{r} 1 \\ + 6 \\ \hline \end{array}$$

n.
$$\begin{array}{r} 3 \\ + 4 \\ \hline \end{array}$$

o.
$$\begin{array}{r} 2 \\ + 4 \\ \hline \end{array}$$

7. Solve the word problems. Draw pictures to help you!
   Think: Are you asked the total? Or do you already know the total?

a. Lisa has three goldfish and Lauren has six. How many goldfish do they have together?	**b.** Paul has seven T-shirts. Two of them are red. How many are not red?
c. A fish bowl has four fish swimming in it. Lisa added four more. How many fish are now in it?	**d.** Paul has nine toy cars. Six of them are in the living room. The rest of them Paul cannot find. How many cars are missing?
e. Jill wants to have hats for all seven of her dolls. She has found three hats so far. How many does she still need?	**f.** Brenda ate two cookies, and later she ate four more. How many cookies did she eat?

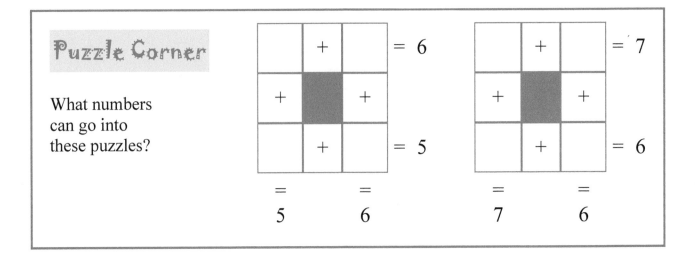

Puzzle Corner

What numbers can go into these puzzles?

44

Sums with 8

1. Here are some different ways to group eight marbles into two groups. Write the additions.

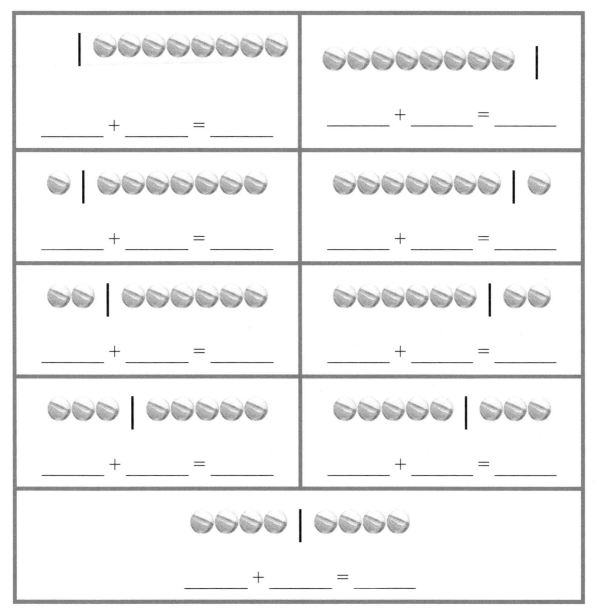

_____ + _____ = _____

_____ + _____ = _____

_____ + _____ = _____

_____ + _____ = _____

_____ + _____ = _____

_____ + _____ = _____

_____ + _____ = _____

_____ + _____ = _____

_____ + _____ = _____

2. **Drill.** Don't write the answers here. Just solve them in your head.

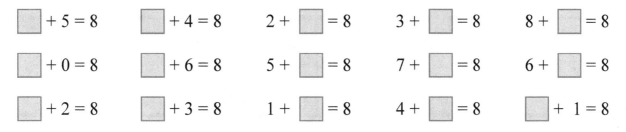

☐ + 5 = 8	☐ + 4 = 8	2 + ☐ = 8	3 + ☐ = 8	8 + ☐ = 8
☐ + 0 = 8	☐ + 6 = 8	5 + ☐ = 8	7 + ☐ = 8	6 + ☐ = 8
☐ + 2 = 8	☐ + 3 = 8	1 + ☐ = 8	4 + ☐ = 8	☐ + 1 = 8

3. Play "8 Out" *and/or* "Some Went Hiding" with 8 objects (see the introduction).

4. Fill in the missing numbers. You may draw dots to help. Notice the patterns!

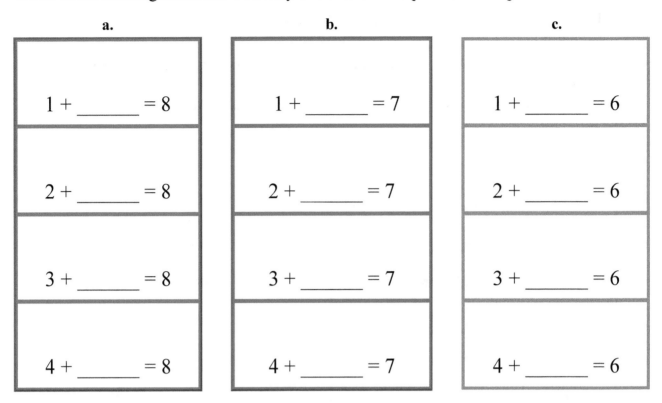

a.

1 + _____ = 8

2 + _____ = 8

3 + _____ = 8

4 + _____ = 8

b.

1 + _____ = 7

2 + _____ = 7

3 + _____ = 7

4 + _____ = 7

c.

1 + _____ = 6

2 + _____ = 6

3 + _____ = 6

4 + _____ = 6

5. Draw the missing marbles. Write the additions.

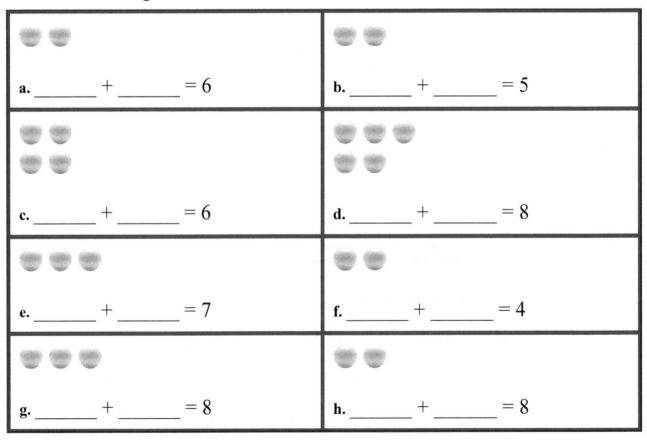

a. _____ + _____ = 6

b. _____ + _____ = 5

c. _____ + _____ = 6

d. _____ + _____ = 8

e. _____ + _____ = 7

f. _____ + _____ = 4

g. _____ + _____ = 8

h. _____ + _____ = 8

6. Find the missing numbers.

a.	b.	c.	d.
$3 + 4 =$ _____	$6 + 2 =$ _____	$6 + 1 =$ _____	$2 + 5 =$ _____
$4 + 4 =$ _____	$5 + 2 =$ _____	$1 + 7 =$ _____	$2 + 6 =$ _____

e.	f.	g.	h.
$5 +$ _____ $= 7$	$4 +$ _____ $= 8$	$3 +$ _____ $= 7$	$2 +$ _____ $= 8$
$5 +$ _____ $= 8$	$4 +$ _____ $= 7$	$3 +$ _____ $= 8$	$2 +$ _____ $= 7$

7. Add.

a.
$$\begin{array}{r} 4 \\ + \ 2 \\ \hline \end{array}$$

b.
$$\begin{array}{r} 6 \\ + \ 2 \\ \hline \end{array}$$

c.
$$\begin{array}{r} 3 \\ + \ 3 \\ \hline \end{array}$$

d.
$$\begin{array}{r} 7 \\ + \ 1 \\ \hline \end{array}$$

e.
$$\begin{array}{r} 5 \\ + \ 2 \\ \hline \end{array}$$

f.
$$\begin{array}{r} 1 \\ + \ 2 \\ \hline \end{array}$$

g.
$$\begin{array}{r} 6 \\ + \ 1 \\ \hline \end{array}$$

h.
$$\begin{array}{r} 4 \\ + \ 3 \\ \hline \end{array}$$

i.
$$\begin{array}{r} 5 \\ + \ 1 \\ \hline \end{array}$$

j.
$$\begin{array}{r} 3 \\ + \ 2 \\ \hline \end{array}$$

8. Which number is greater? Or are they equal? Write < , > or = .
   (Write one of the alligator mouths or the equal sign).

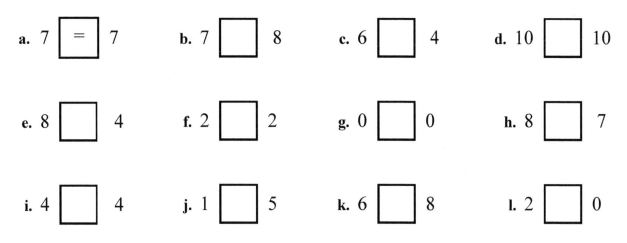

a. 7 ☐=☐ 7   b. 7 ☐ 8   c. 6 ☐ 4   d. 10 ☐ 10

e. 8 ☐ 4   f. 2 ☐ 2   g. 0 ☐ 0   h. 8 ☐ 7

i. 4 ☐ 4   j. 1 ☐ 5   k. 6 ☐ 8   l. 2 ☐ 0

Adding Many Numbers

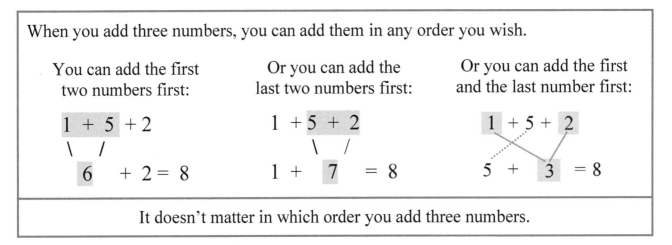

When you add three numbers, you can add them in any order you wish.

You can add the first two numbers first:	Or you can add the last two numbers first:	Or you can add the first and the last number first:
1 + 5 + 2	1 + 5 + 2	1 + 5 + 2
6 + 2 = 8	1 + 7 = 8	5 + 3 = 8

It doesn't matter in which order you add three numbers.

1. Add in different orders. Which way is easier for you?

a.	**b.**	**c.**
Add the last two numbers first.	Add the first two numbers first.	Add the last two numbers first.
2 + 5 + 2 = _____	3 + 1 + 5 = _____	2 + 5 + 3 = _____
Add the first and last number first.	Add the first and last number first.	Add the first two numbers first.
2 + 5 + 2 = _____	3 + 1 + 5 = _____	2 + 5 + 3 = _____

2. Add. Again, you can add in any order.

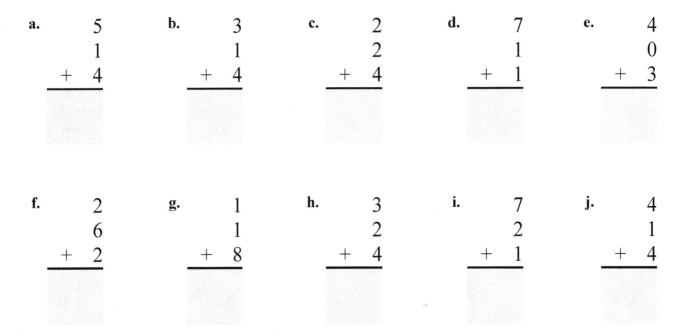

a.
```
    5
    1
+   4
_____
```

b.
```
    3
    1
+   4
_____
```

c.
```
    2
    2
+   4
_____
```

d.
```
    7
    1
+   1
_____
```

e.
```
    4
    0
+   3
_____
```

f.
```
    2
    6
+   2
_____
```

g.
```
    1
    1
+   8
_____
```

h.
```
    3
    2
+   4
_____
```

i.
```
    7
    2
+   1
_____
```

j.
```
    4
    1
+   4
_____
```

3. Solve. You can draw pictures to help.

a. Molly was picking flowers. First she picked two pretty ones.
Then she found some more and picked three more flowers.
Then she picked two more.
How many flowers does Molly have now?

b. Emily put three chairs in a row. Behind them she put another
three chairs, and yet behind them three more chairs.
Draw a picture.
How many chairs did she use?

c. Jack has 10 rabbits. One morning when he came to see them,
he only saw 6 rabbits. How many were missing?

4. Are these additions right? Circle <u>true</u> or <u>false</u>.

a. $1 + 2 + 3 = 6$ true *or* false	**b.** $2 + 2 + 3 = 8$ true *or* false
$1 + 7 + 2 = 9$ true *or* false	$2 + 5 + 2 = 9$ true *or* false

5. Match the addition problems to the right pictures and solve them.

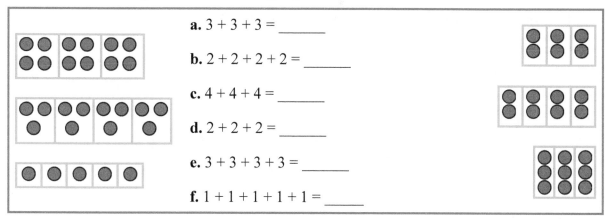

a. $3 + 3 + 3 =$ _____

b. $2 + 2 + 2 + 2 =$ _____

c. $4 + 4 + 4 =$ _____

d. $2 + 2 + 2 =$ _____

e. $3 + 3 + 3 + 3 =$ _____

f. $1 + 1 + 1 + 1 + 1 =$ _____

6. Write the additions that match the number line jumps.

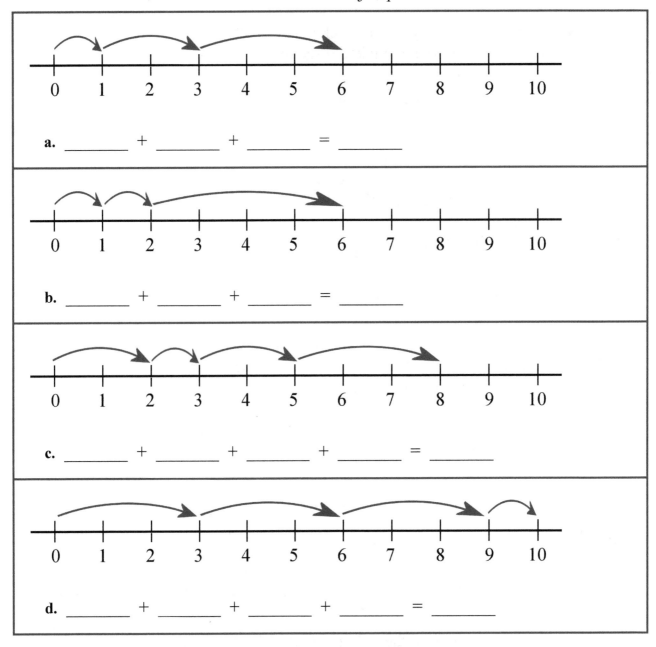

a. _____ + _____ + _____ = _____

b. _____ + _____ + _____ = _____

c. _____ + _____ + _____ + _____ = _____

d. _____ + _____ + _____ + _____ = _____

7. Add four numbers. You can color the numbers you want to add first!

a.	b.	c.
$1 + 2 + 2 + 3 =$ ____	$4 + 0 + 3 + 2 =$ ____	$2 + 5 + 3 + 0 =$ ____
$5 + 0 + 1 + 2 =$ ____	$3 + 1 + 2 + 1 =$ ____	$1 + 1 + 2 + 1 =$ ____
$2 + 1 + 3 + 4 =$ ____	$7 + 1 + 1 + 1 =$ ____	$2 + 1 + 5 + 2 =$ ____

Addition Practice 2

1. Add.

a. $4 + 4 =$ _____	**b.** $4 + 3 =$ _____	**c.** $2 + 4 =$ _____
$6 + 2 =$ _____	$5 + 2 =$ _____	$1 + 6 =$ _____

2. **Double** means two times the same thing! Draw dots or sticks. Write the total in the box.

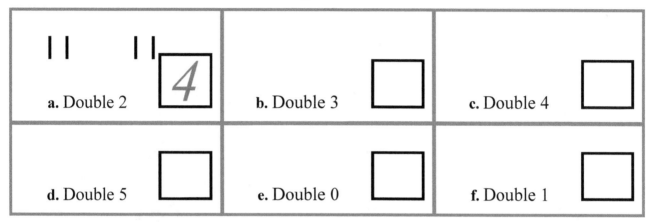

a. Double 2	**b.** Double 3	**c.** Double 4
d. Double 5	**e.** Double 0	**f.** Double 1

3. Draw jumps for each of the additions. Find the answer.

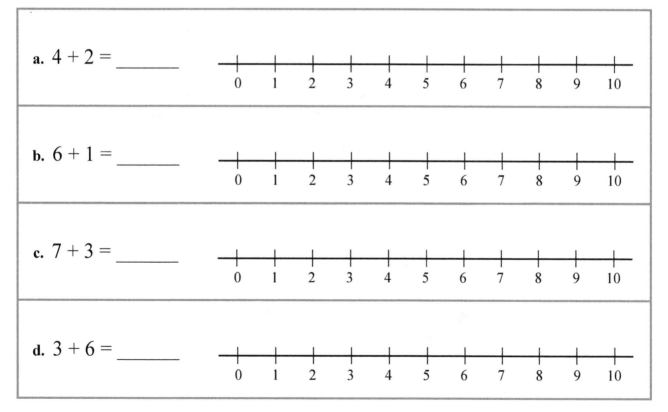

a. $4 + 2 =$ _____

b. $6 + 1 =$ _____

c. $7 + 3 =$ _____

d. $3 + 6 =$ _____

4. You can add the numbers in either order! Which way is easier?

a.	b.	c.	d.
7 + 2 = _____	2 + 5 = _____	6 + 2 = _____	1 + 4 = _____
2 + 7 = _____	5 + 2 = _____	2 + 6 = _____	4 + 1 = _____

5. Let's make charts! In the first chart, add one each time. In the second, add two each time. In the third, add three each time.

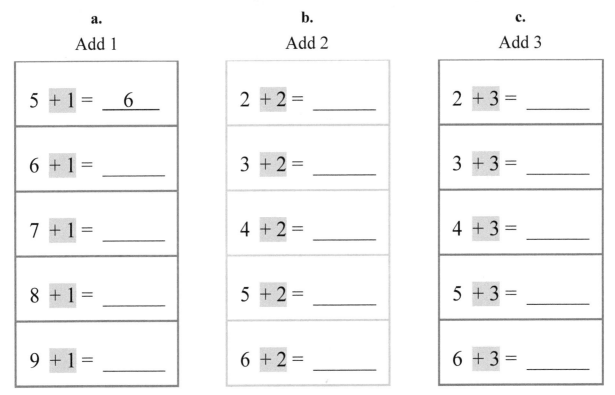

a. Add 1	b. Add 2	c. Add 3
5 + 1 = __6__	2 + 2 = _____	2 + 3 = _____
6 + 1 = _____	3 + 2 = _____	3 + 3 = _____
7 + 1 = _____	4 + 2 = _____	4 + 3 = _____
8 + 1 = _____	5 + 2 = _____	5 + 3 = _____
9 + 1 = _____	6 + 2 = _____	6 + 3 = _____

6. Fill in the addition tables. Add the number above and the number to the left.

+	1	2	3
1			
2		4	
3			

+	1	2	3
4			
5			
6			

Sums with 9

1. Here are some different ways to group nine marbles into two groups. Write the addition sentences.

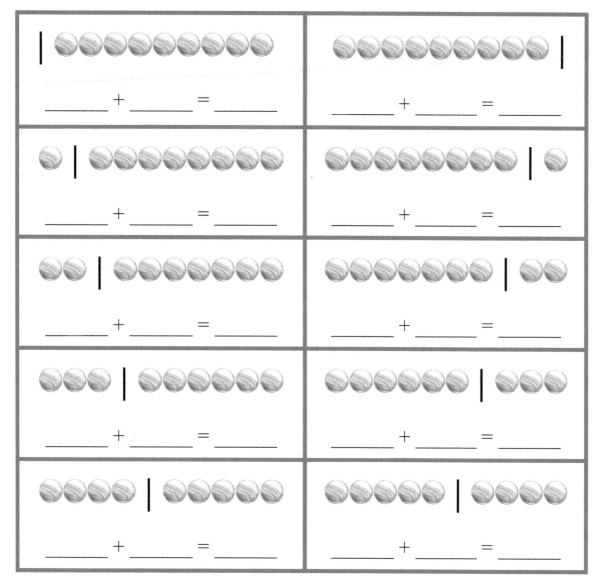

2. **Drill.** Don't write the answers here. Just solve the answers in your head.

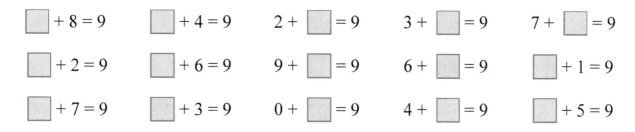

$\boxed{} + 8 = 9$ $\boxed{} + 4 = 9$ $2 + \boxed{} = 9$ $3 + \boxed{} = 9$ $7 + \boxed{} = 9$

$\boxed{} + 2 = 9$ $\boxed{} + 6 = 9$ $9 + \boxed{} = 9$ $6 + \boxed{} = 9$ $\boxed{} + 1 = 9$

$\boxed{} + 7 = 9$ $\boxed{} + 3 = 9$ $0 + \boxed{} = 9$ $4 + \boxed{} = 9$ $\boxed{} + 5 = 9$

3. Play "9 Out" *and/or* "Some Went Hiding" with 9 objects (see the introduction).

4. Fill in the missing numbers.

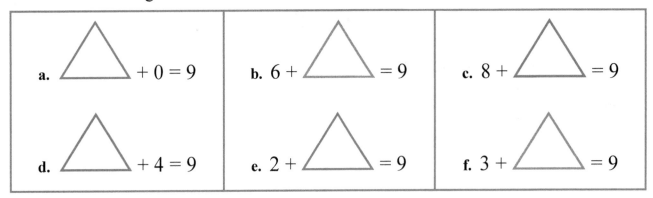

a. △ + 0 = 9

b. 6 + △ = 9

c. 8 + △ = 9

d. △ + 4 = 9

e. 2 + △ = 9

f. 3 + △ = 9

5. Fill in the missing numbers. You may draw dots to help. Notice the patterns!

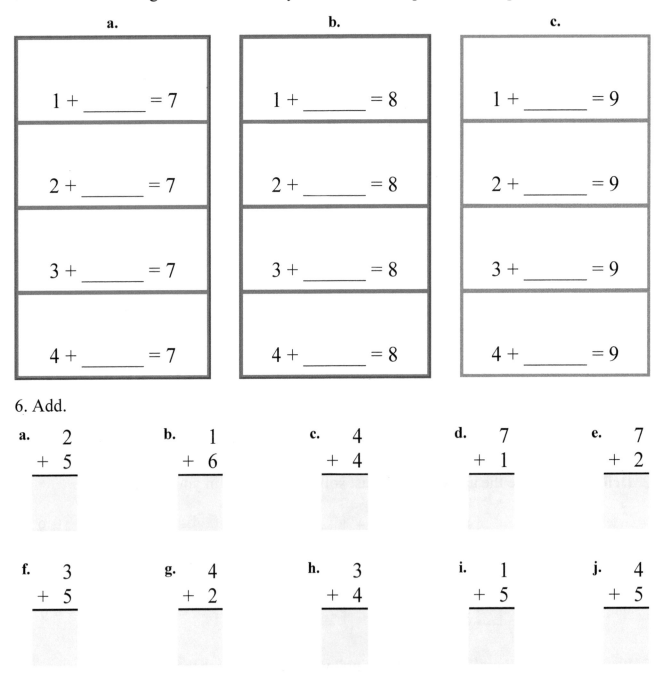

a.

1 + _____ = 7

2 + _____ = 7

3 + _____ = 7

4 + _____ = 7

b.

1 + _____ = 8

2 + _____ = 8

3 + _____ = 8

4 + _____ = 8

c.

1 + _____ = 9

2 + _____ = 9

3 + _____ = 9

4 + _____ = 9

6. Add.

a. 2
   + 5
   ____

b. 1
   + 6
   ____

c. 4
   + 4
   ____

d. 7
   + 1
   ____

e. 7
   + 2
   ____

f. 3
   + 5
   ____

g. 4
   + 2
   ____

h. 3
   + 4
   ____

i. 1
   + 5
   ____

j. 4
   + 5
   ____

7. Solve the word problems. Write an addition sentence or a "missing addend" sentence for each problem. Think: "Is it asking the total? Or, do I already know the total, and something else is being asked?" You can draw a picture to help!

a. Mom has two eggs at home. The cake recipe calls for five eggs. How many more eggs will she need?	**b.** You see four crayons in the crayon box and the rest of them are missing. The full box has eight crayons. How many crayons are missing?
c. Jenny and Penny each have five goldfish. How many do they have together? Betty has three goldfish. How many do the three girls have together?	**d.** You have two dollars. Can you buy a doll for nine dollars? Father has eight dollars. How much money do you have together? Can you buy the doll together?
e. There are two red chairs in the living room and six red chairs in the kitchen, and none in the other rooms. How many red chairs are in the house?	**f.** Joshua has $5. He wants to buy a truck for $7. How many more dollars will he need?
g. If you have $8, and a gift for Mom costs $10, how much more money do you need?	**h.** Jack bought nails for five dollars and screws for four dollars. How much money did he spend in all?

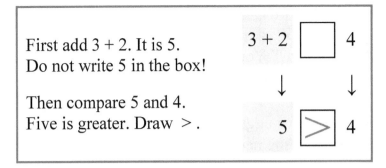

First add 3 + 2. It is 5.
Do not write 5 in the box!

$3 + 2$ ☐ 4

↓   ↓

Then compare 5 and 4.
Five is greater. Draw >.

5 $\boxed{>}$ 4

8. First add. Write the answer below (not in the box!). Then write < , > or = .

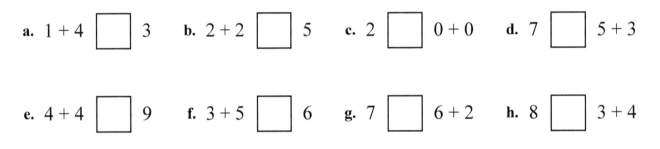

a. $5 + 2$ ☐ 4

↓   ↓

☐ 4

b. $4 + 4$ ☐ 7

↓   ↓

☐ 7

c. 2 ☐ $1 + 1$

↓   ↓

☐

d. 7 ☐ $3 + 6$

↓   ↓

☐

9. Add in your head. Compare the sum to the other number. Then write < or > .

a. $1 + 4$ ☐ 3   **b.** $2 + 2$ ☐ 5   **c.** 2 ☐ $0 + 0$ **d.** 7 ☐ $5 + 3$

e. $4 + 4$ ☐ 9   **f.** $3 + 5$ ☐ 6   **g.** 7 ☐ $6 + 2$ **h.** 8 ☐ $3 + 4$

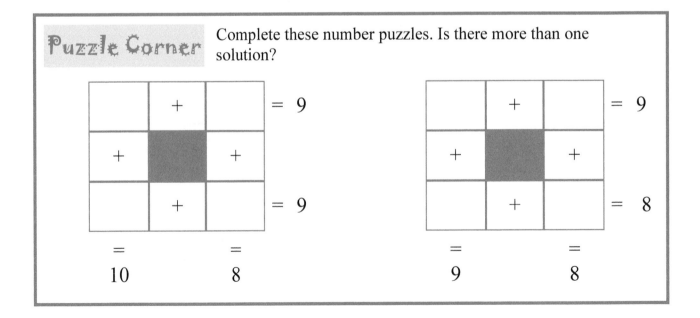

Puzzle Corner   Complete these number puzzles. Is there more than one solution?

Sums with 10

1. Here are some different ways to group ten marbles into two groups. Write the additions.

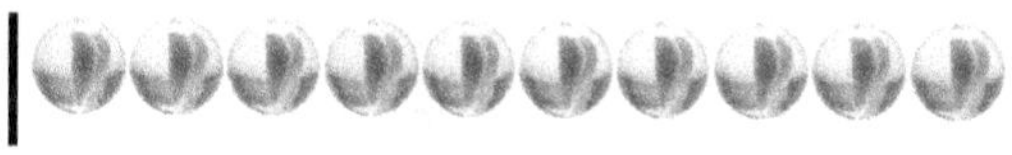

_____ + _____ = _____

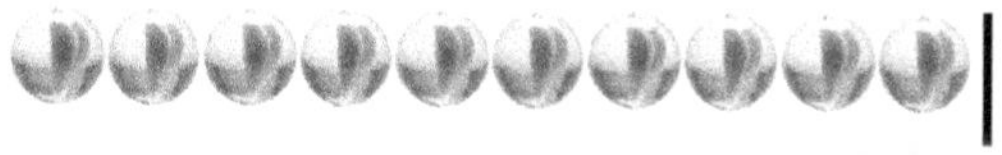

_____ + _____ = _____

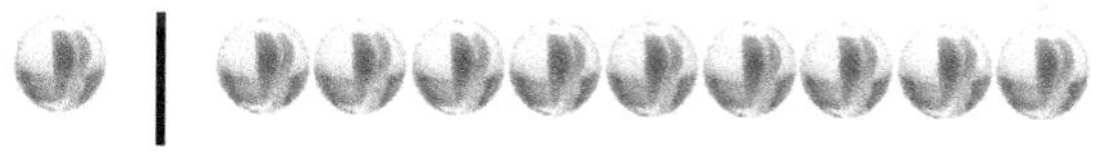

_____ + _____ = _____

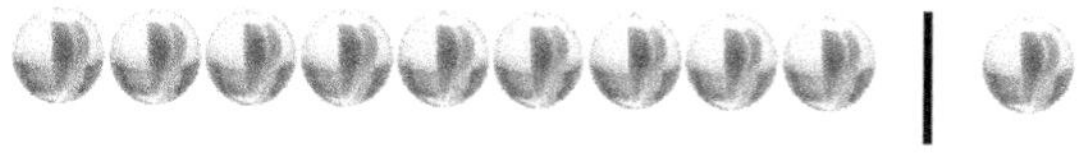

_____ + _____ = _____

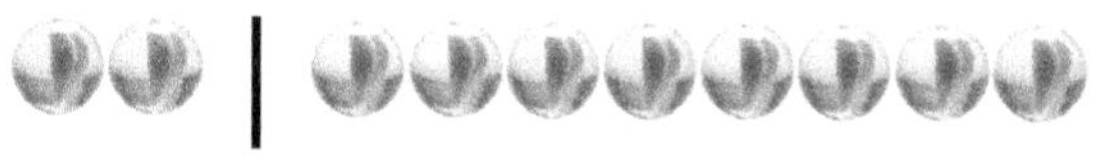

_____ + _____ = _____

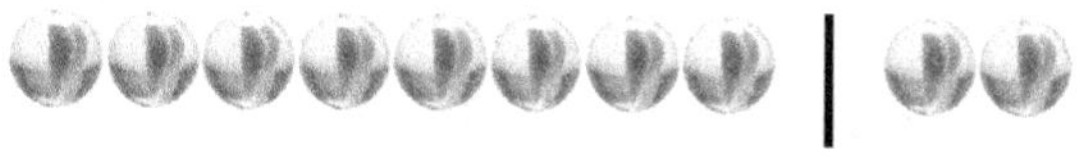

_____ + _____ = _____

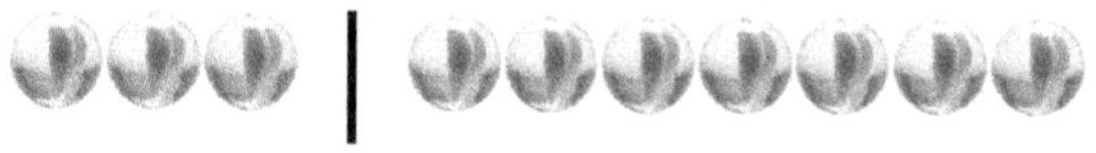

_____ + _____ = _____

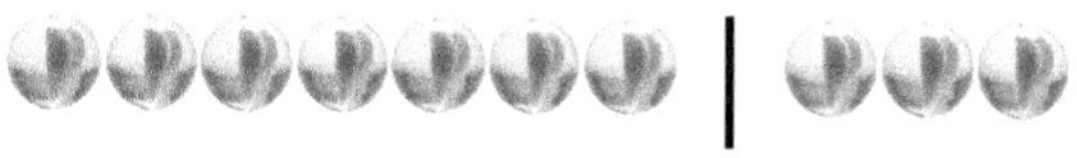

_____ + _____ = _____

_____ + _____ = _____

_____ + _____ = _____

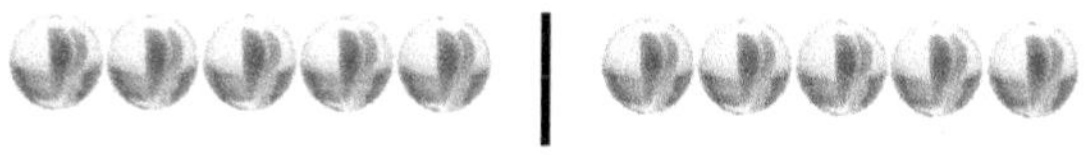

_____ + _____ = _____

2. Play "10 Out" *and/or* "Some Went Hiding" with 10 objects (see the introduction).

3. **Drill.** Do not write the answers here. Just think of the answers in your head.

$\square + 6 = 10$ $\square + 4 = 10$ $1 + \square = 10$ $6 + \square = 10$ $3 + \square = 10$

$\square + 3 = 10$ $\square + 5 = 10$ $7 + \square = 10$ $9 + \square = 10$ $4 + \square = 10$

$\square + 8 = 10$ $\square + 9 = 10$ $2 + \square = 10$ $5 + \square = 10$ $8 + \square = 10$

4. Fill in the missing numbers. You may draw dots to help. Notice the patterns!

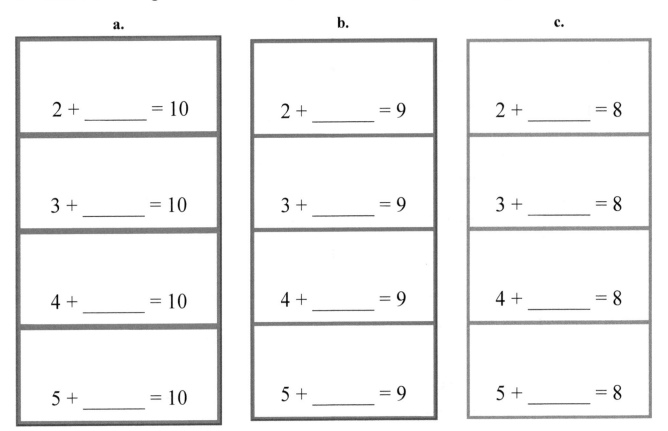

a.	b.	c.
$2 + \underline{\hspace{1cm}} = 10$	$2 + \underline{\hspace{1cm}} = 9$	$2 + \underline{\hspace{1cm}} = 8$
$3 + \underline{\hspace{1cm}} = 10$	$3 + \underline{\hspace{1cm}} = 9$	$3 + \underline{\hspace{1cm}} = 8$
$4 + \underline{\hspace{1cm}} = 10$	$4 + \underline{\hspace{1cm}} = 9$	$4 + \underline{\hspace{1cm}} = 8$
$5 + \underline{\hspace{1cm}} = 10$	$5 + \underline{\hspace{1cm}} = 9$	$5 + \underline{\hspace{1cm}} = 8$

5. Connect two numbers together if they make ten.

4	1	6	4	4	3
5	2	3	8	2	2
3	7	1	9	2	5
3	6	5	7	5	7
9	0	3	2	3	8

6. Which number is greater? Or are they equal? Write < , > or = .
   (Write one of the alligator mouths or the equal sign).

   a. 6 ☐ 7   **b.** 10 ☐ 8   **c.** 6 ☐ 8   **d.** 10 ☐ 10

   e. 8 ☐ 6   **f.** 5 ☐ 5   **g.** 9 ☐ 8   **h.** 5 ☐ 10

7. First add. Think of the answers in your head. Then compare and write < , > or = .

a. $1 + 9$ ☐ 9	**b.** $4 + 4$ ☐ 9	**c.** 6 ☐ $5 + 2$	**d.** 9 ☐ $5 + 4$
e. $5 + 5$ ☐ 10	**f.** $3 + 5$ ☐ 7	**g.** 10 ☐ $6 + 3$	**h.** 7 ☐ $7 + 1$

8. Which numbers add up to ten? Fill in the missing numbers.

a. ☐ $+ 10 = 10$	**b.** $6 +$ ☐ $= 10$	**c.** ☐ $+ 3 = 10$
☐ $+ 5 = 10$	$2 +$ ☐ $= 10$	☐ $+ 8 = 10$
☐ $+ 1 = 10$	$4 +$ ☐ $= 10$	☐ $+ 9 = 10$

9. Draw a line to the correct answer.

7	$7 + 1$	**8**
	$2 + 6$	
	$3 + 4$	
	$5 + 2$	
	$4 + 4$	
	$1 + 6$	
	$5 + 3$	

9	$7 + 3$	**10**
	$3 + 6$	
	$4 + 6$	
	$1 + 8$	
	$5 + 4$	
	$3 + 7$	
	$2 + 8$	

10. Solve the word problems.

a. There were three birds in the tree. Seven more flew in. How many birds are now in the tree?	**b.** Tina has seven books from the library. She has read three. How many books has she not read?
c. Jessica has ten dolls. She sees four of them in her room. How many are somewhere else?	**d.** Larry has three toy cars and his brother also has three. How many do they have together?
e. Bill has ten toy cars but he can find only six. How many are missing?	**f.** Jack saw two birds on the lawn and five on the fence. How many birds did he see in all?
g. Together, Jessica and Jenny have ten books. Jenny has five of them. How many does Jessica have?	**h.** The store has ten dolls. Two of them are on the bottom shelf. The rest are on the top shelf. How many dolls are on the top shelf?

Comparisons

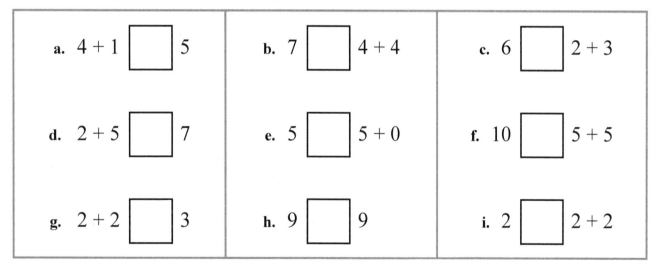 7 = 7 Seven equals seven.	6 = 2 + 4 Six equals two plus four.	= means "EQUALS"
7 < 8 Seven is less than eight.	3 + 4 > 5 Three plus four is greater than five.	< means "IS LESS THAN" > means "IS GREATER THAN"

1. First add in your head. Then compare and write < , > or = .

a. 4 + 1 ☐ 5	**b.** 7 ☐ 4 + 4	**c.** 6 ☐ 2 + 3
d. 2 + 5 ☐ 7	**e.** 5 ☐ 5 + 0	**f.** 10 ☐ 5 + 5
g. 2 + 2 ☐ 3	**h.** 9 ☐ 9	**i.** 2 ☐ 2 + 2

2. Pick a number to write on the line so the comparison is true.

a. 5 6 7 _____ < 6	**b.** 4 5 6 _____ < 5	**c.** 5 6 7 _____ > 6	**d.** 2 3 4 _____ > 3
e. 9 7 5 _____ > 7	**f.** 3 6 9 _____ < 5	**g.** 1 3 7 _____ > 6	**h.** 2 4 6 _____ < 3

3. Pick a number to write on the line so the comparison is true.

a. 2 3 4 2 + _____ = 6	**b.** 4 5 6 1 + _____ < 6	**c.** 1 2 3 4 + _____ < 7
d. 4 5 6 2 + _____ > 6	**e.** 4 5 6 1 + _____ = 6	**f.** 7 8 9 1 + _____ > 9
g. 6 7 8 10 = 2 + _____	**h.** 2 4 6 3 + _____ < 7	**i.** 4 5 6 4 + _____ > 8

4. Compare. Write < , > or = .

a. $4 + 3$ ☐ 5	**b.** $7 + 1$ ☐ 9	**c.** 4 ☐ $4 + 2$
d. $2 + 5$ ☐ 8	**e.** $3 + 4$ ☐ 6	**f.** 6 ☐ $3 + 3$
g. $8 + 2$ ☐ 10	**h.** $9 + 2$ ☐ 9	**i.** 2 ☐ $2 + 1$

5. Challenges! First add in your head. Then write < , > or = .

a. $7 + 3$ ☐ $2 + 8$ **b.** $1 + 1$ ☐ $1 + 4$ **c.** 4 ☐ $1 + 4$

d. $5 + 4$ ☐ $4 + 5$ **e.** $2 + 5$ ☐ $2 + 2$ **f.** 3 ☐ $3 + 1$

g. $2 + 4$ ☐ $2 + 1$ **h.** $10 + 0$ ☐ $0 + 10$ **i.** 0 ☐ $0 + 0$

6. Are these additions right? Circle true or false.

a. $7 + 3 \boxed{} = 10$ true *or* false	**d.** $7 \boxed{} = 1 + 5$ true *or* false
b. $9 \ = \ 5 + 5$ true *or* false	**e.** $2 + 2 \ = \ 1 + 3$ true *or* false
c. $2 + 4 \ = \ 7$ true *or* false	**f.** $3 + 5 \ = \ 7 + 2$ true *or* false

7. What numbers make 10? Draw arrows to illustrate the additions on the number line.

a. $10 = $ _____ $+$ _____

b. $10 = $ _____ $+$ _____

c. $10 = $ _____ $+$ _____

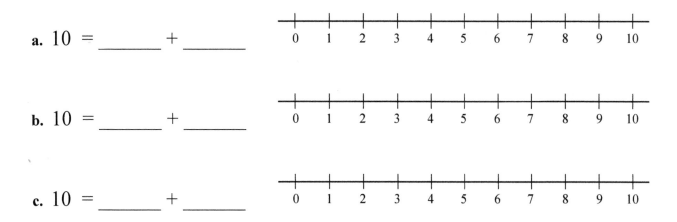

8. Fill in as much of the addition table as you can, and do not worry about the rest.
   Color the square blue if the answer is 8.

+	1	2	3	4	5	6	7
0							
1							
2							
3							
4							
5							

Review of Addition Facts

1. Write different sums of 5 and sums of 6.

5 = _____ + _____	6 = _____ + _____
5 = _____ + _____	6 = _____ + _____
5 = _____ + _____	6 = _____ + _____
5 = _____ + _____	6 = _____ + _____

2. Draw a line to the correct answer.

3. Find the missing addends

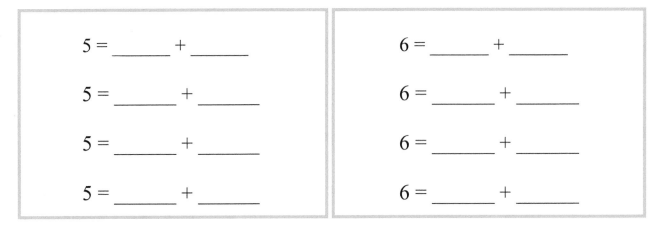

```
          4 + 1
          2 + 3
          3 + 3
          5 + 0
    5     4 + 2     6
          5 + 1
          0 + 6
          1 + 4
          2 + 4
```

_____ + 2 = 6   _____ + 0 = 6

2 + _____ = 5   0 + _____ = 5

1 + _____ = 5   3 + _____ = 6

6 + _____ = 6   4 + _____ = 6

_____ + 1 = 6   _____ + 4 = 5

4. Compare. Write < , > , or = .

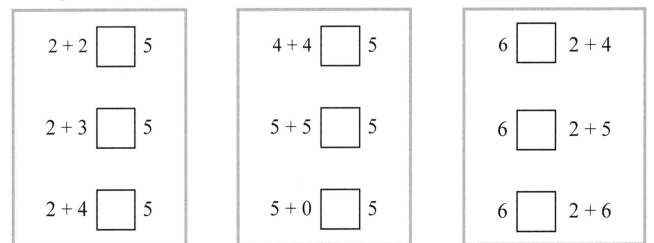

2 + 2 ☐ 5	4 + 4 ☐ 5	6 ☐ 2 + 4
2 + 3 ☐ 5	5 + 5 ☐ 5	6 ☐ 2 + 5
2 + 4 ☐ 5	5 + 0 ☐ 5	6 ☐ 2 + 6

5. Write different sums of 7 and sums of 8.

7 = _____ + _____   8 = _____ + _____

7 = _____ + _____   8 = _____ + _____

7 = _____ + _____   8 = _____ + _____

7 = _____ + _____   8 = _____ + _____

7 = _____ + _____   8 = _____ + _____

7 = _____ + _____   8 = _____ + _____

6. Draw a line to the correct answer.

7   4 + 3   8
      2 + 6
      3 + 5
      4 + 4
      5 + 2
      1 + 6
      5 + 3
      7 + 1
      6 + 2

7. Find the missing addends

_____ + 2 = 7   _____ + 4 = 8

_____ + 4 = 7   3 + _____ = 7

2 + _____ = 8   3 + _____ = 8

5 + _____ = 8   7 + _____ = 8

6 + _____ = 7   5 + _____ = 7

8. Compare. Write < , > , or = .

3 + 3 ☐ 7   6 + 1 ☐ 7   8 ☐ 6 + 4

4 + 3 ☐ 7   6 + 6 ☐ 7   8 ☐ 4 + 4

5 + 3 ☐ 7   6 + 4 ☐ 7   8 ☐ 5 + 4

9. Write different sums of 9 and sums of 10.

9 = _____ + _____   10 = _____ + _____

9 = _____ + _____   10 = _____ + _____

9 = _____ + _____   10 = _____ + _____

9 = _____ + _____   10 = _____ + _____

9 = _____ + _____   10 = _____ + _____

9 = _____ + _____   10 = _____ + _____

10. Draw a line to the correct answer.

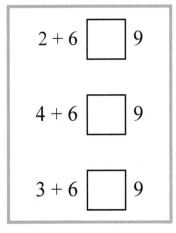

	2 + 7
	3 + 6
	4 + 6
	5 + 5
9	9 + 1	10
	1 + 8
	5 + 4
	3 + 7
	2 + 8

11. Find the missing addends

_____ + 2 = 10   _____ + 6 = 9

_____ + 4 = 9   7 + _____ = 10

2 + _____ = 9   3 + _____ = 9

5 + _____ = 10   7 + _____ = 9

6 + _____ = 10   4 + _____ = 10

12. Compare. Write < , > , or = .

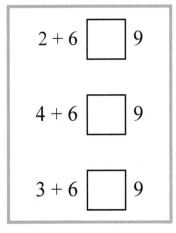

2 + 6 ☐ 9

4 + 6 ☐ 9

3 + 6 ☐ 9

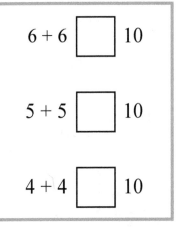

6 + 6 ☐ 10

5 + 5 ☐ 10

4 + 4 ☐ 10

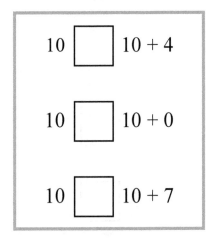

10 ☐ 10 + 4

10 ☐ 10 + 0

10 ☐ 10 + 7

13. Add.

a.	b.	c.
8 + 1 = _____	4 + 1 + 1 = _____	5 + 2 + 0 + 0 = _____
6 + 2 = _____	8 + 2 + 0 = _____	4 + 3 + 1 + 2 = _____
1 + 7 = _____	1 + 3 + 6 = _____	1 + 2 + 2 + 1 = _____
3 + 4 = _____	2 + 2 + 4 = _____	2 + 3 + 1 + 3 = _____

14. Fill in as much as you can of the addition table.

+	2	4	3	6	7	5	8
1							
3							
4							
2							

Puzzle Corner △ represents a number, and ☐ represents another number. Solve what they are in each case (a, b, and c).
Hint: Make a guess! Then check if your guess is correct.
If not, change your guess.

a.	b.	c.
△ + △ = 6	☐ + △ = 7	☐ + △ + △ = 7
☐ + △ = 8	☐ + ☐ = 10	☐ + △ = 5

Chapter 2: Subtraction Within 0-10
Introduction

The second chapter of *Math Mammoth Grade 1-A* covers the concept of subtraction, the relationship between addition and subtraction, and the various meanings of subtraction. Keep in mind that the specific lessons in the chapter can take several days to finish. They are not "daily lessons."

In the first lesson, *Subtraction Is Taking Away*, the child learns the basic meaning of subtraction as taking away objects, and learns to write subtractions from an illustration where some objects are crossed out. The child can figure out the answers by simply counting how many objects are left.

If the child does not yet know the word "minus," it is a good idea to introduce it first orally. Use blocks or other concrete objects. For example, show the child eight blocks and take away three blocks. Then use both kinds of wordings: "Eight blocks, take away three blocks, leaves five blocks. Eight blocks *minus* three blocks *equals* five blocks." Then let the child do the same. Play with concrete objects until the child can use the words "minus" and "equals" in his or her own speech.

In the next lesson, the child counts down to subtract, which ties in subtraction with the number line. This is a transitional strategy to solve subtraction problems, because later students will learn more efficient ways to subtract, but it is important conceptually. For now, the student can solve $9 - 3$ by counting down three steps from nine: eight, seven, six. So the answer is six.

The following lesson, *Subtraction and Addition in the Same Picture*, begins the study of the relationship between addition and subtraction. This concept will span several lessons. This first lesson presents two sets of objects, such as blue and white balls, and the student writes both an addition sentence and a subtraction sentence from this illustration.

The lesson *When Can You Subtract?* concentrates on the idea that some subtractions, such as $4 - 5$, are meaningless when you think of taking away. The child also makes subtraction patterns in this lesson.

Then we continue studying the connection between addition and subtraction in the lesson *Two Subtractions from One Addition*. As an example, the child writes both $8 - 3 = 5$ and $8 - 5 = 3$ from the addition $3 + 5 = 8$. This idea ties in with fact families, a concept that is coming up soon.

In the lesson *Two Parts—One Total,* we study word problems that do not involve the idea of taking away but have two parts making up a total. For example, if there are 10 flowers of which some are white and some are red, and seven of them are white, how many are red? We know the "parts" (the red and white flowers) add up to 10, so we can write a missing addend sentence $7 + \_\_ = 10$. This can be solved by subtracting $10 - 7$ or by knowing the addition fact $7 + 3 = 10$.

Then we study fact families. This means writing two additions and two subtractions using the same three numbers. Fact families will be used extensively in the next chapter.

In the lesson *How Many More?* students find how many more or how many fewer objects one person has than the other by drawing the objects. This lesson can easily be done with manipulatives if desired.

In the very next lesson, *"How Many More" Problems and Differences,* we continue the theme, this time writing a missing addend addition for problems that ask "how many more." For example, Veronica has 4 marbles and Ann has 6. We write the missing addend sentence $4 + \_\_\_ = 6$ to find how how many more Ann has. In the next lesson the child then learns to write subtraction sentences for such problems.

The Lessons in Chapter 2

Helpful Resources on the Internet

Use these free online resources to supplement the "bookwork" as you see fit.

> You can also access this list of links at **https://links.mathmammoth.com/gr1ch2**

DISCLAIMER: *We check these links a few times a year. However, we cannot guarantee that the links have not changed. Parental supervision is always recommended.*

Basic Addition & Subtraction Facts — online practice
An ad-free online program to practice basic addition and subtraction at MathMammoth.com website. Also works as an offline program in most browsers.
https://www.mathmammoth.com/practice/addition-single-digit.php

Hidden Picture Subtraction Game
Click on the correct answer to each subtraction problem and uncover a hidden picture.
https://www.mathmammoth.com/practice/mystery-picture-subtraction#min=0&max=11

Kids' Subtraction Quiz from Mr. Martini's Classroom
A five-question quiz. Choose the maximum number used from the list of numbers below the quiz.
http://www.thegreatmartinicompany.com/Math-Quick-Quiz/subtraction-kid-quiz.html

Subtraction Harvest
Choose the correct answers for the subtraction problems to harvest the apples.
https://www.sheppardsoftware.com/mathgames/earlymath/subHarvest.htm

Butterfly Ten Frames Game
Practice subtracting from 10 with this interactive online activity.
https://coolsciencelab.com/butterfly_ten_frames.html

Busy Bees
Figure out how many of the 10 bees went inside the hive.
http://www.hbschool.com/activity/busy_bees/index.html

Soccer Subtraction
Click to make the players disappear until the subtraction sentence is true.
https://www.ictgames.com/soccer_subtraction.html

Pearl Search
Click on the clam that contains the correct answer to the subtraction problem, and collect a pearl for each correct answer. The sooner you get the pearls, the higher your score!
https://www.sheppardsoftware.com/mathgames/popup/popup_subtraction.htm

Fruit Shoot Number-Line Subtraction
Click on the subtraction sentence on the fruit that matches the number line. Choose level 3 for this game.
https://www.sheppardsoftware.com/mathgames/earlymath/FS_NumberLine_minus.htm

Mathemorphosis Subtraction Game
Solve simple subtraction problems to help the caterpillar transform into a butterfly.
https://mrnussbaum.com/mathemorphosis-online-game/

Subtraction Sense–Make Subtraction Sentences
Drag and drop the number cards to make 'sum' sense. Try to answer 8 questions in 2 minutes!
https://primarygames.co.uk/pg2/sumsense/sumsub.html

Sea Life Subtraction
Solve subtraction problems while creating your own beautiful underwater scene.
https://www.free-training-tutorial.com/subtraction/sealife/sl-subtraction.html

Snowy's Friend
Practice subtraction facts while helping Snowy through many different levels, and gather pieces to help make a new snowman friend.
https://fun4thebrain.com/subtraction/snowyfriend.html

Subtraction Word Problem Quiz
Eight simple word problems for first grade.
https://www.math4children.com/games-k-to-6/1st%20grade/subtraction%20word%20problems/index.html

Block Buster
Click on blocks with numbers that form fact families. The blocks must be touching each other.
https://www.roomrecess.com/mobile/BlockBuster/play.html

Basic Fact Families
Find the missing equation in each fact family.
https://mrnussbaum.com/1-oa-b-3-grade-1-common-core-fact-families

Subtraction 4-in-a-Row
Play the 4-in-a-row game as you answer subtraction problems.
https://www.multiplication.com/games/play/subtraction-4-row

Balancing Calculations
Answer as many questions as you can in this timed online quiz.
http://www.snappymaths.com/mixed/addsubrelate/interactive/addsubbalancew10/addsubbalancew10.htm

Tux Math
A versatile free software for math facts with many options. Includes all operations.
https://sourceforge.net/projects/tuxmath/

Subtraction Is "Taking Away"

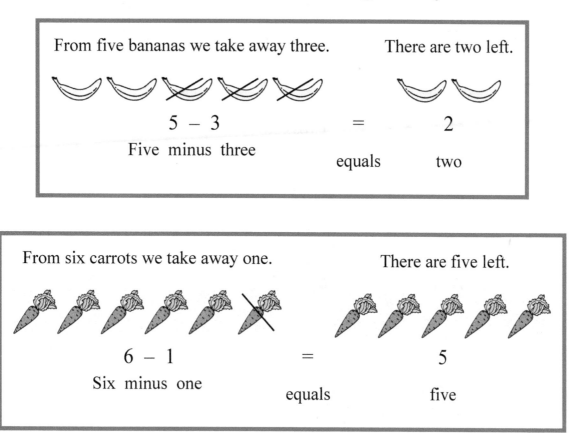

From five bananas we take away three.   There are two left.

5 – 3   =   2

Five minus three

equals   two

From six carrots we take away one.   There are five left.

6 – 1   =   5

Six minus one

equals   five

1. Cross out objects. How many are left? Read each subtraction sentence aloud using the words "minus" and "equals".

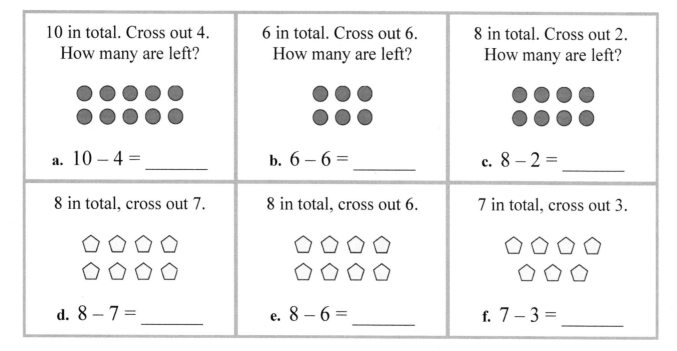

10 in total. Cross out 4. How many are left?	6 in total. Cross out 6. How many are left?	8 in total. Cross out 2. How many are left?
a. 10 – 4 = _____	**b.** 6 – 6 = _____	**c.** 8 – 2 = _____
8 in total, cross out 7.	8 in total, cross out 6.	7 in total, cross out 3.
d. 8 – 7 = _____	**e.** 8 – 6 = _____	**f.** 7 – 3 = _____

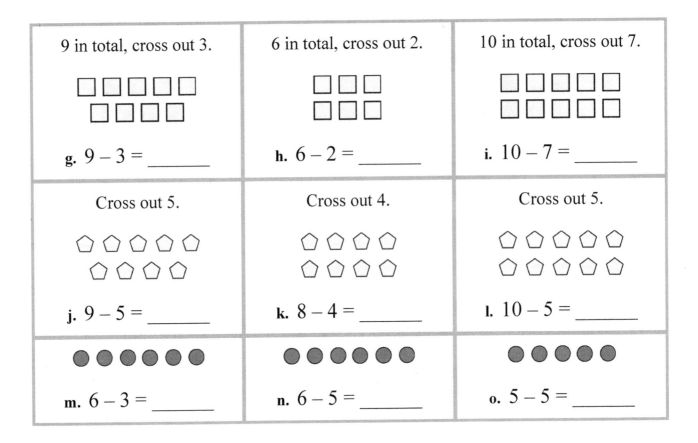

9 in total, cross out 3. g. $9 - 3 =$ _____	6 in total, cross out 2. h. $6 - 2 =$ _____	10 in total, cross out 7. i. $10 - 7 =$ _____
Cross out 5. j. $9 - 5 =$ _____	Cross out 4. k. $8 - 4 =$ _____	Cross out 5. l. $10 - 5 =$ _____
m. $6 - 3 =$ _____	n. $6 - 5 =$ _____	o. $5 - 5 =$ _____

2. Subtract. Cover the crossed-out objects with your finger to see how many are left.
   Read each sentence using the words "minus" and "equals".

a. $5 - 1 =$ _____	b. $6 - 3 =$ _____	c. $5 - 3 =$ _____
d. $4 - 1 =$ _____	e. $5 - 2 =$ _____	f. $6 - 2 =$ _____
g. $9 - 3 =$ _____	h. $7 - 1 =$ _____	i. $7 - 2 =$ _____
j. $8 - 2 =$ _____	k. $7 - 3 =$ _____	l. $8 - 4 =$ _____

3. Draw small circles to illustrate the numbers and cross out some of them to match the subtraction problem.

a. $8 - 3 =$ _____	b. $5 - 1 =$ _____	c. $10 - 7 =$ _____
d. $7 - 2 =$ _____	e. $10 - 1 =$ _____	f. $9 - 7 =$ _____
g. $6 - 3 =$ _____	h. $7 - 3 =$ _____	i. $10 - 4 =$ _____
j. $9 - 5 =$ _____	k. $10 - 6 =$ _____	l. $6 - 4 =$ _____

4. Write a subtraction sentence to match the picture.

a. _____ − _____ = _____	b. _____ − _____ = _____
c. _____ − _____ = _____	d. _____ − _____ = _____
e. _____ − _____ = _____	f. _____ − _____ = _____

Count Down to Subtract

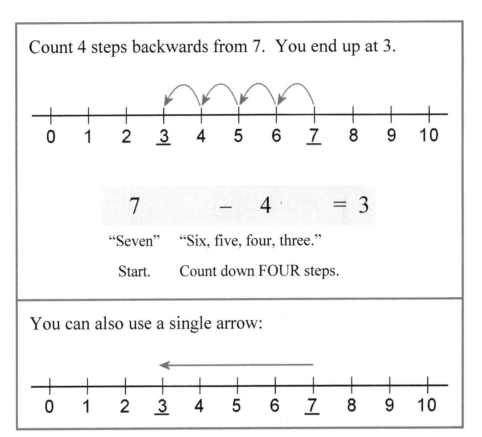

Count 4 steps backwards from 7. You end up at 3.

| 0   1   2   <u>3</u>   4   5   6   <u>7</u>   8   9   10 |

| 7 | – | 4 | = 3 |

"Seven"   "Six, five, four, three."

Start.   Count down FOUR steps.

You can also use a single arrow:

| 0   1   2   <u>3</u>   4   5   6   <u>7</u>   8   9   10 |

1. Draw steps (or a single arrow) to illustrate the subtraction sentence.

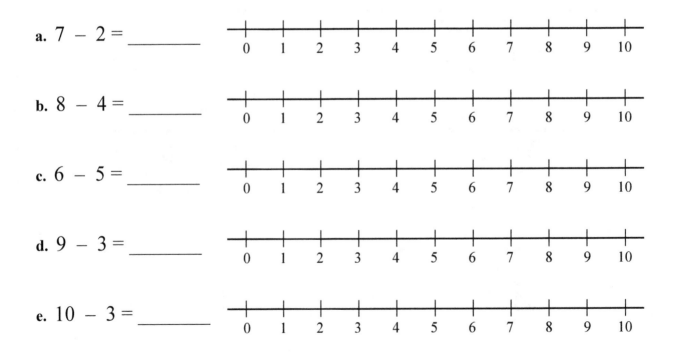

a. $7 - 2 =$ _____

      0   1   2   3   4   5   6   7   8   9   10

b. $8 - 4 =$ _____

      0   1   2   3   4   5   6   7   8   9   10

c. $6 - 5 =$ _____

      0   1   2   3   4   5   6   7   8   9   10

d. $9 - 3 =$ _____

      0   1   2   3   4   5   6   7   8   9   10

e. $10 - 3 =$ _____

      0   1   2   3   4   5   6   7   8   9   10

2. Write the subtraction sentence that the arrows illustrate.

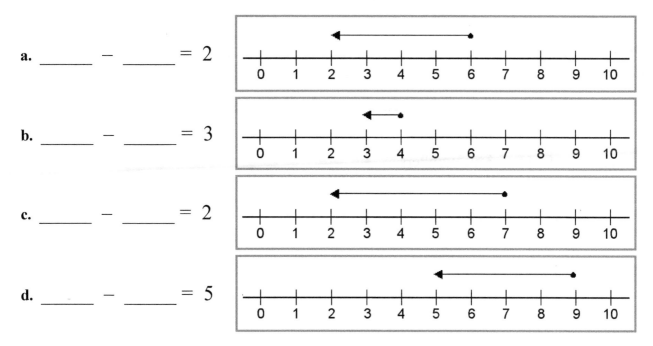

a. _____ – _____ = 2

b. _____ – _____ = 3

c. _____ – _____ = 2

d. _____ – _____ = 5

3. Draw an arrow for the subtraction sentence and solve.

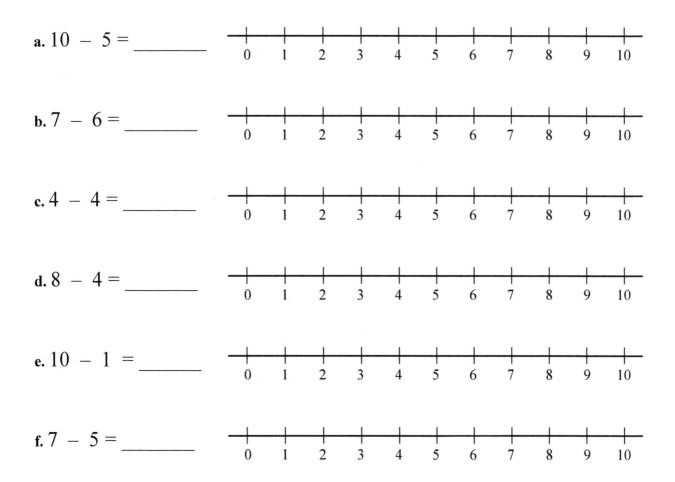

a. $10 - 5 =$ _____

b. $7 - 6 =$ _____

c. $4 - 4 =$ _____

d. $8 - 4 =$ _____

e. $10 - 1 =$ _____

f. $7 - 5 =$ _____

75

4. Write the previous and the next numbers.

a. _____ , 5 , _____   b. _____ , 2 , _____   c. _____ , 8 , _____

d. _____ , 6 , _____   e. _____ , 4 , _____   f. _____ , 9 , _____

5. Write the two previous numbers.

a. _____ , _____ , 7   b. _____ , _____ , 4   c. _____ , _____ , 10

d. _____ , _____ , 6   e. _____ , _____ , 2   f. _____ , _____ , 8

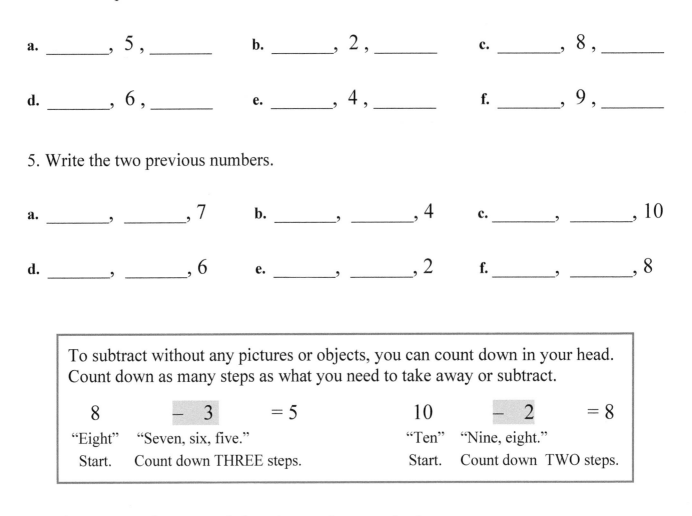

To subtract without any pictures or objects, you can count down in your head.
Count down as many steps as what you need to take away or subtract.

8   – 3   = 5   10   – 2   = 8
"Eight"   "Seven, six, five."   "Ten"   "Nine, eight."
Start.   Count down THREE steps.   Start.   Count down TWO steps.

6. Subtract one. The answer is just the previous number!

a.	b.	c.	d.
6 – 1 = _____	5 – 1 = _____	9 – 1 = _____	4 – 1 = _____
8 – 1 = _____	7 – 1 = _____	3 – 1 = _____	10 – 1 = _____

7. Subtract 2 or 3. You can count down. Compare the problems.

a.	b.	c.	d.
6 – 2 = _____	9 – 2 = _____	7 – 2 = _____	10 – 2 = _____
6 – 3 = _____	9 – 3 = _____	7 – 3 = _____	10 – 3 = _____

8. Solve the word problems. Write a subtraction sentence for each.

a. There were 7 birds in a tree. Three flew away. How many are left?	**b.** Mom has 10 silver plates in the cupboard. She took out four. How many are still in the cupboard?
c. All 9 girls in the class were jumping rope. Then four of them left. How many kept on jumping?	**d.** Josh took five of his 10 toy cars to a friend's house. How many cars did he leave at home?
e. Of her eight puzzles, Fanny put three in the closet. How many were left to play with?	**f.** Tina had 6 bunches of flowers. She sold six of them. How many were left?

9. Do these problems if you know the numbers past 10. The number line will help.

0 1 2 3 4 5 6 7 8 9 10 11 12 13 14 15 16 17

a. $14 - 2 =$ _____ $14 - 4 =$ _____	**b.** $16 - 1 =$ _____ $16 - 3 =$ _____	**c.** $11 - 3 =$ _____ $11 - 4 =$ _____
d. $17 - 2 =$ _____ $17 - 3 =$ _____	**e.** $12 - 1 =$ _____ $12 - 2 =$ _____	**f.** $13 - 2 =$ _____ $13 - 3 =$ _____

Subtraction and Addition in the Same Picture

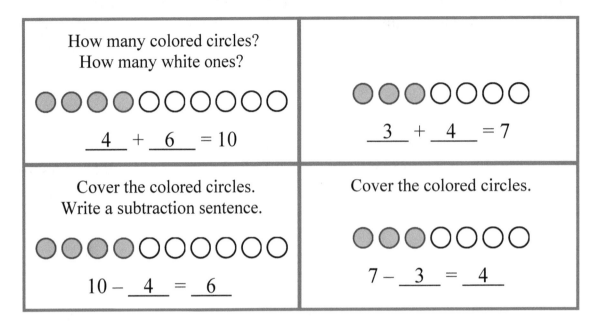

How many colored circles?
How many white ones?

<u>4</u> + <u>6</u> = 10

<u>3</u> + <u>4</u> = 7

Cover the colored circles.
Write a subtraction sentence.

10 – <u>4</u> = <u>6</u>

Cover the colored circles.

7 – <u>3</u> = <u>4</u>

1. Make an addition sentence and a subtraction sentence from the same picture.

a.

_____ + _____ = _____

7 – _____ = _____

b.

_____ + _____ = _____

6 – _____ = _____

c.

_____ + _____ = _____

5 – _____ = _____

d.

_____ + _____ = _____

6 – _____ = _____

e.

_____ + _____ = _____

8 – _____ = _____

f.

_____ + _____ = _____

6 – _____ = _____

2. Make an addition sentence and a subtraction sentence for the same picture.

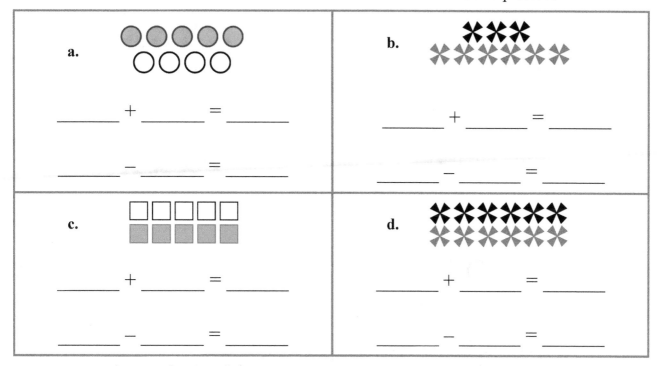

a.

_____ + _____ = _____

_____ − _____ = _____

b.

_____ + _____ = _____

_____ − _____ = _____

c.

_____ + _____ = _____

_____ − _____ = _____

d.

_____ + _____ = _____

_____ − _____ = _____

3. In each problem, draw circles and then color them to fit the addition sentence.
   Then cover the **COLORED** circles and make a subtraction sentence.

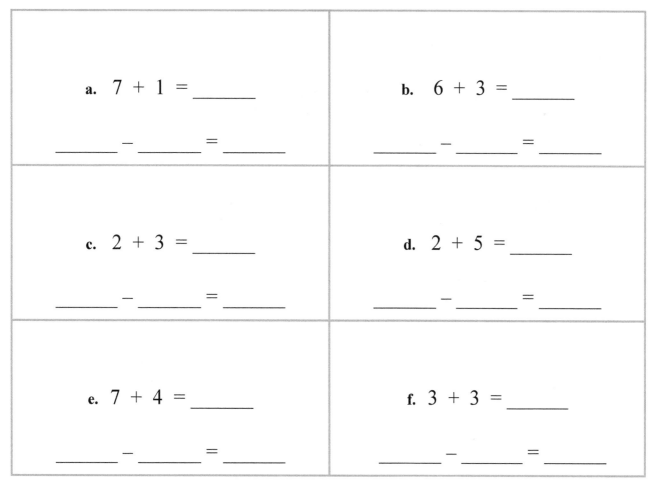

a. 7 + 1 = _____

_____ − _____ = _____

b. 6 + 3 = _____

_____ − _____ = _____

c. 2 + 3 = _____

_____ − _____ = _____

d. 2 + 5 = _____

_____ − _____ = _____

e. 7 + 4 = _____

_____ − _____ = _____

f. 3 + 3 = _____

_____ − _____ = _____

79

4. Cover the colored objects, and write a subtraction sentence to fit the picture.

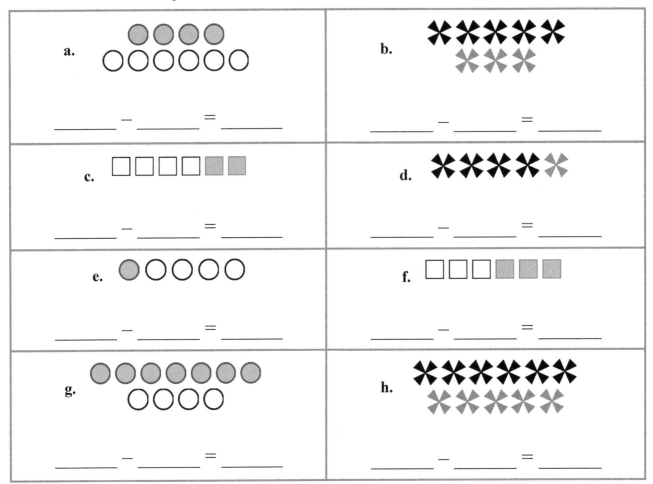

5. In each problem, draw some circles and color some circles to fit the addition sentence. Then cover the **COLORED** circles and make a subtraction sentence.

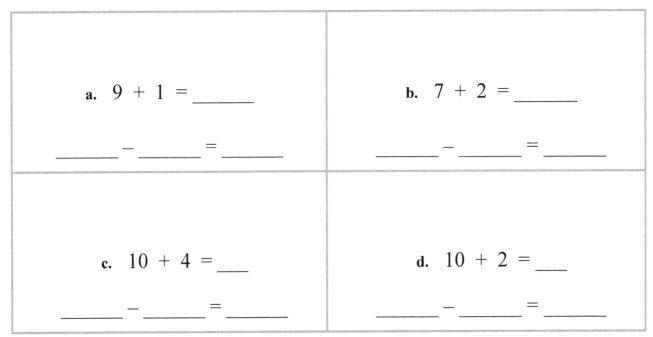

6. Draw circles to fit the subtraction sentence. Write an addition sentence, too.

a. 9 − 4 = ___

_____ + _____ = _____

b. 10 − 5 = ___

_____ + _____ = _____

c. 8 − 5 = ___

_____ + _____ = _____

d. 8 − 4 = ___

_____ + _____ = _____

e. 7 − 4 = ___

_____ + _____ = _____

f. 9 − 8 = ___

_____ + _____ = _____

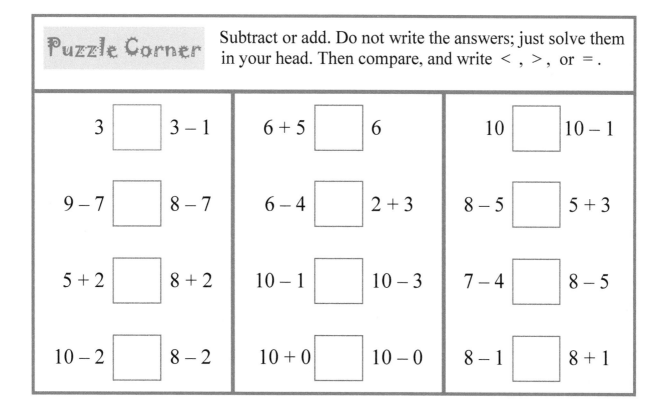

Puzzle Corner Subtract or add. Do not write the answers; just solve them in your head. Then compare, and write < , > , or = .

3 ☐ 3 − 1	6 + 5 ☐ 6	10 ☐ 10 − 1
9 − 7 ☐ 8 − 7	6 − 4 ☐ 2 + 3	8 − 5 ☐ 5 + 3
5 + 2 ☐ 8 + 2	10 − 1 ☐ 10 − 3	7 − 4 ☐ 8 − 5
10 − 2 ☐ 8 − 2	10 + 0 ☐ 10 − 0	8 − 1 ☐ 8 + 1

When Can You Subtract?

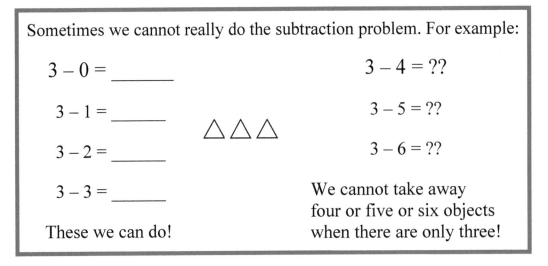

Sometimes we cannot really do the subtraction problem. For example:

3 – 0 = _____

3 – 1 = _____

3 – 2 = _____

△ △ △

3 – 3 = _____

These we can do!

3 – 4 = ??

3 – 5 = ??

3 – 6 = ??

We cannot take away
four or five or six objects
when there are only three!

1. Write the subtraction problems you *can* do when there are...

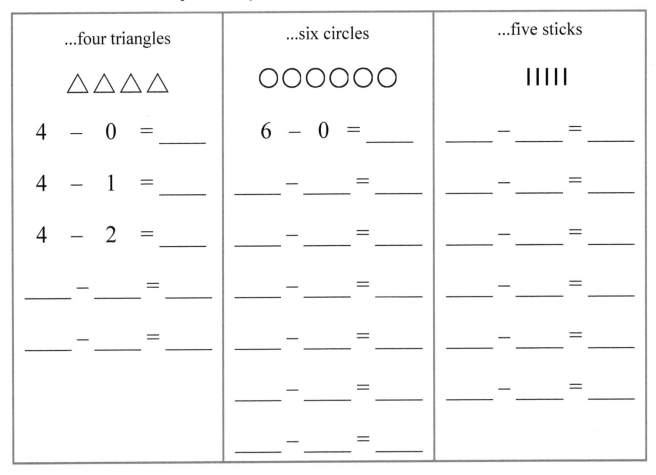

...four triangles

△ △ △ △

4 – 0 = ____

4 – 1 = ____

4 – 2 = ____

____ – ____ = ____

____ – ____ = ____

...six circles

○ ○ ○ ○ ○ ○

6 – 0 = ____

____ – ____ = ____

____ – ____ = ____

____ – ____ = ____

____ – ____ = ____

____ – ____ = ____

...five sticks

| | | | |

____ – ____ = ____

____ – ____ = ____

____ – ____ = ____

____ – ____ = ____

____ – ____ = ____

You cannot do a subtraction problem in whole numbers (0, 1, 2, 3, ...)
when the second number is _____ than the first number.

2. Count down to subtract.

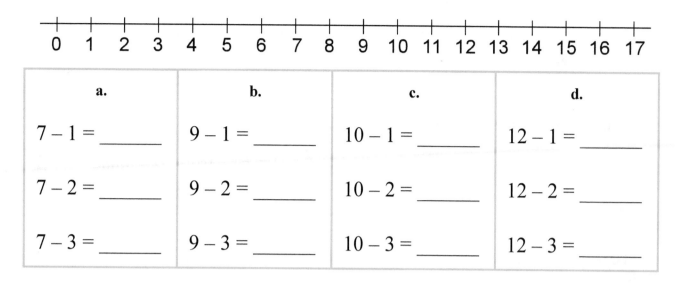

a.	b.	c.	d.
7 − 1 = _____	9 − 1 = _____	10 − 1 = _____	12 − 1 = _____
7 − 2 = _____	9 − 2 = _____	10 − 2 = _____	12 − 2 = _____
7 − 3 = _____	9 − 3 = _____	10 − 3 = _____	12 − 3 = _____

3. Continue the patterns as long as you can!

a.	b.	c.
7 − 0 = ___	10 − 5 = ___	8 − 2 = ___
7 − 1 = ___	9 − 5 = ___	7 − 2 = ___
7 − 2 = ___	8 − 5 = ___	6 − 2 = ___
7 − ___ = ___	___ − 5 = ___	___ − ___ = ___
___ − ___ = ___	___ − ___ = ___	___ − ___ = ___
___ − ___ = ___	___ − ___ = ___	___ − ___ = ___
___ − ___ = ___	___ − ___ = ___	___ − ___ = ___
___ − ___ = ___	___ − ___ = ___	___ − ___ = ___
___ − ___ = ___	___ − ___ = ___	___ − ___ = ___

4. Find the subtractions where you can't take away that many, and cross them out.
   You *don't* have to write the answers!

   $4 - 0$ $7 - 7$ $5 - 6$ $3 - 6$ $4 - 4$ $3 - 10$

   $4 - 5$ $7 - 9$ $10 - 1$ $3 - 4$ $2 - 4$ $4 - 3$

5. Are these subtractions right? Circle true or false.

a. $7 - 1 = 8$ true *or* false	**d.** $5 - 2 = 6$ true *or* false
b. $9 - 2 = 7$ true *or* false	**e.** $10 - 8 = 1$ true *or* false
c. $10 - 5 = 4$ true *or* false	**f.** $6 - 3 = 3$ true *or* false

6. For each problem, answer these questions: Can the child buy the item? Yes or no.
   If yes, how much money will she/he have left?
   If not, how much more money would she/he need to buy the item?

a. Jennie has three dollars. She wants to buy a doll that costs five dollars.	**b.** Jessie has $5. He wants a ball that costs $2.
c. Lola has seven dollars. She wants to buy a Lego set that costs four dollars.	**d.** Marvin has $5. He wants a book that costs $6.
e. Jack has eight dollars. He wants to buy a construction set that costs ten dollars.	**f.** Mary has seven dollars. She wants a car that costs two dollars.
g. Faye has $12. She wants to buy a game that costs $4.	**h.** Anthony has ten dollars. He wants a game that costs fifteen dollars.

7. Subtraction can be written this way too! Write the answer below the line.

a. 10
  − 3
  ———

b.   8
  − 7
  ———

c.   6
  − 5
  ———

d.   8
  − 6
  ———

e.   8
  − 0
  ———

f.   7
  − 7
  ———

g.   7
  − 6
  ———

h.   6
  − 6
  ———

i.   6
  − 1
  ———

j.   9
  − 4
  ———

k. 10
  − 8
  ———

l.   4
  − 0
  ———

m.   6
  − 4
  ———

n.   7
  − 2
  ———

o.   9
  − 3
  ———

8. If the answer is four, color the box red. If the answer is five, color the box orange. If the answer is ten, color the box yellow. And, if you can't subtract, color the box light blue.

$10 - 6$	$6 - 7$	$5 - 6$	$5 - 10$	$7 - 3$
$1 - 6$	$10 - 5$	$9 - 10$	$8 - 3$	$1 - 3$
$7 - 9$	$1 - 2$	$10 - 0$	$3 - 5$	$6 - 9$
$4 - 8$	$6 - 1$	$0 - 9$	$9 - 4$	$7 - 8$
$8 - 4$	$0 - 4$	$2 - 6$	$6 - 9$	$9 - 5$

Two Subtractions from One Addition

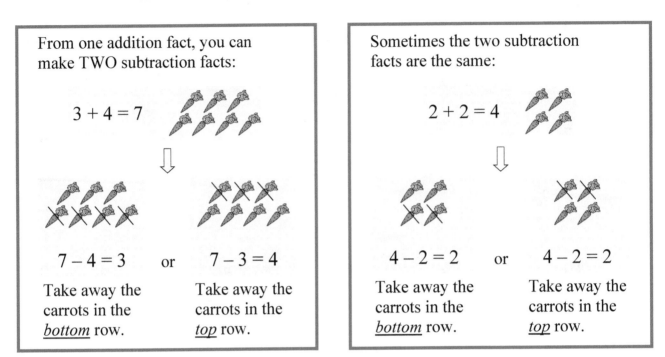

From one addition fact, you can make TWO subtraction facts:

$3 + 4 = 7$

⇓

$7 - 4 = 3$ or $7 - 3 = 4$

Take away the carrots in the _bottom_ row.

Take away the carrots in the _top_ row.

Sometimes the two subtraction facts are the same:

$2 + 2 = 4$

⇓

$4 - 2 = 2$ or $4 - 2 = 2$

Take away the carrots in the _bottom_ row.

Take away the carrots in the _top_ row.

1. Write one addition and two subtraction sentences. First subtract the things in the bottom row then the ones in the top row.

a. $1 + 3 = \underline{\quad 4 \quad}$

$4 - 3 = \underline{\qquad}$

or $4 - 1 = \underline{\qquad}$

b. $2 + 3 = 5$

$5 - \underline{\qquad} = \underline{\qquad}$

or $5 - \underline{\qquad} = \underline{\qquad}$

c. $\underline{\qquad} + \underline{\qquad} = \underline{\qquad}$

$\underline{\qquad} - \underline{\qquad} = \underline{\qquad}$

or $\underline{\qquad} - \underline{\qquad} = \underline{\qquad}$

d. $\underline{\qquad} + \underline{\qquad} = \underline{\qquad}$

$\underline{\qquad} - \underline{\qquad} = \underline{\qquad}$

or $\underline{\qquad} - \underline{\qquad} = \underline{\qquad}$

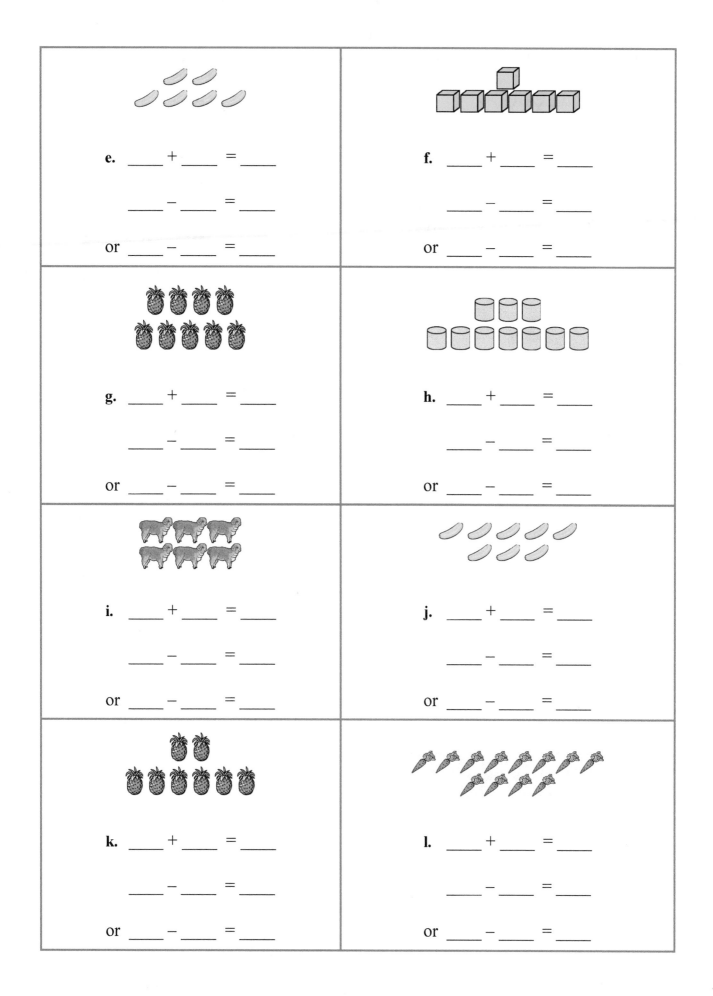

e. ____ + ____ = ____

    ____ – ____ = ____

or ____ – ____ = ____

f. ____ + ____ = ____

    ____ – ____ = ____

or ____ – ____ = ____

g. ____ + ____ = ____

    ____ – ____ = ____

or ____ – ____ = ____

h. ____ + ____ = ____

    ____ – ____ = ____

or ____ – ____ = ____

i. ____ + ____ = ____

    ____ – ____ = ____

or ____ – ____ = ____

j. ____ + ____ = ____

    ____ – ____ = ____

or ____ – ____ = ____

k. ____ + ____ = ____

    ____ – ____ = ____

or ____ – ____ = ____

l. ____ + ____ = ____

    ____ – ____ = ____

or ____ – ____ = ____

2. Complete the addition fact and the subtraction facts. Draw Xs in two groups.

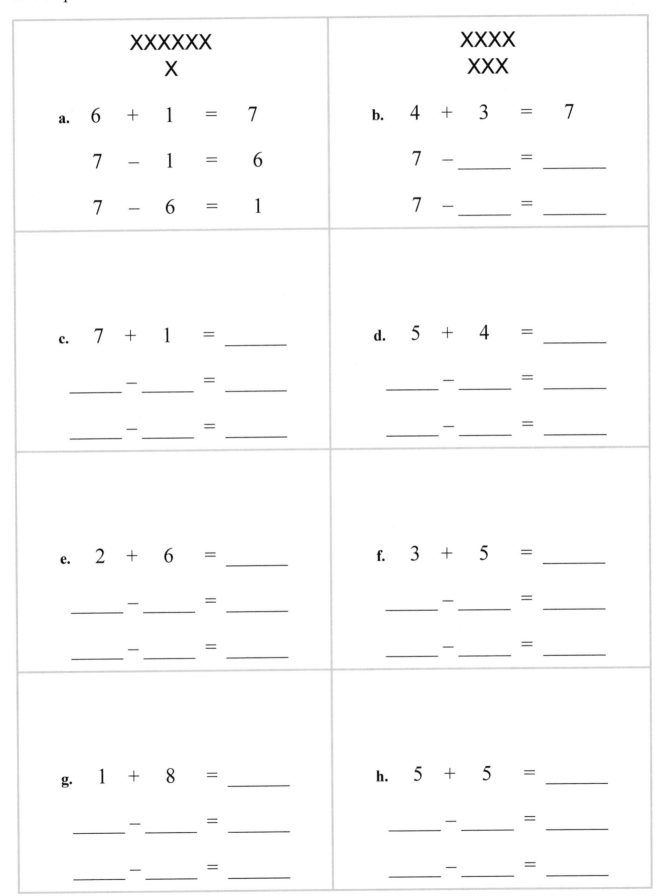

XXXXXX
X

a. 6 + 1 = 7

7 – 1 = 6

7 – 6 = 1

XXXX
XXX

b. 4 + 3 = 7

7 – ____ = ____

7 – ____ = ____

c. 7 + 1 = ____

____ – ____ = ____

____ – ____ = ____

d. 5 + 4 = ____

____ – ____ = ____

____ – ____ = ____

e. 2 + 6 = ____

____ – ____ = ____

____ – ____ = ____

f. 3 + 5 = ____

____ – ____ = ____

____ – ____ = ____

g. 1 + 8 = ____

____ – ____ = ____

____ – ____ = ____

h. 5 + 5 = ____

____ – ____ = ____

____ – ____ = ____

Two Parts — One Total

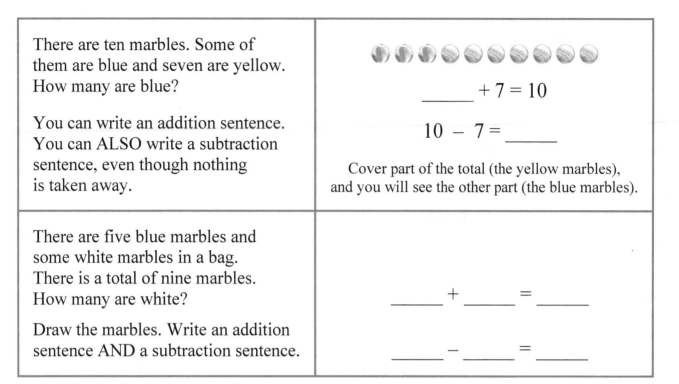

There are ten marbles. Some of them are blue and seven are yellow. How many are blue?

You can write an addition sentence. You can ALSO write a subtraction sentence, even though nothing is taken away.

_____ + 7 = 10

10 − 7 = _____

Cover part of the total (the yellow marbles), and you will see the other part (the blue marbles).

There are five blue marbles and some white marbles in a bag. There is a total of nine marbles. How many are white?

Draw the marbles. Write an addition sentence AND a subtraction sentence.

_____ + _____ = _____

_____ − _____ = _____

1. Solve the word problems. Write an addition sentence AND a subtraction sentence.

a. Mother put some blue and some red flowers in a vase. Jen counted five red ones and a total of ten. How many of the flowers are blue?

_____ + _____ = _____

_____ − _____ = _____

b. There are nine children on a team, and four of them are boys. How many are girls?

_____ + _____ = _____

_____ − _____ = _____

c. Jack has ten socks in his basket. Eight of them are white, and the rest are black. How many are black?

_____ + _____ = _____

_____ − _____ = _____

d. Mary saw eight chairs on the lawn, and two had blown over. How many were still standing upright?

_____ + _____ = _____

_____ − _____ = _____

2. For each picture, make a word problem that is solved by subtraction.

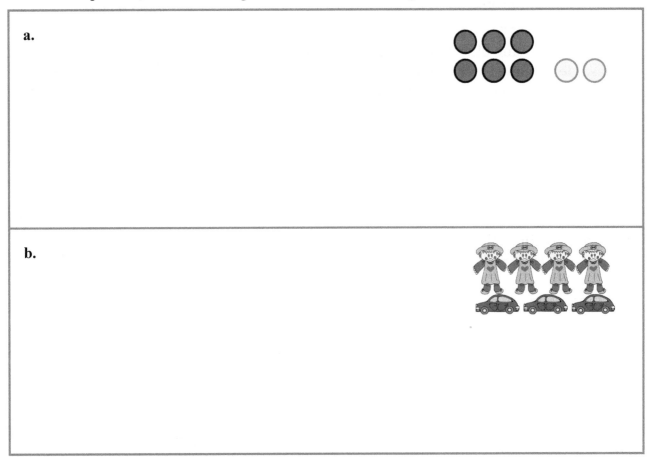

a.

b.

3. Write an addition sentence for the pictures.

a. _____ + _____ + _____ = _____

b. _____ + _____ + _____ = _____

c. _____ + _____ + _____ = _____

d. _____ + _____ + _____ = _____

4. Draw the missing marbles to match the addition sentence.

a. 3 + 2 + _____ = 8

b. 1 + 5 + _____ = 10

5. Draw a picture to solve these problems.

a. Jane had some red, some blue, and some yellow
   roses in a vase. Two of the roses were blue, and
   two were red. If she had a total of ten roses,
   how many of them were yellow?

b. Seven birds sat in a tree. One of
   them was black, two were blue,
   and the rest were brown.
   How many were brown?

c. Mary has two long pencils and two medium-
   sized ones. The rest of her pencils are short
   If she owns nine pencils in all, how many
   of her pencils are short?

Fact Families

Two addition facts and two subtraction facts form a <u>fact family</u> if they use the same three numbers.

For example, from 5, 3, and 2 we get the fact family on the right:

5

●● / ●●●

$2 + 3 = 5$ $5 - 3 = 2$

$3 + 2 = 5$ $5 - 2 = 3$

1. Write the fact families that match the pictures.

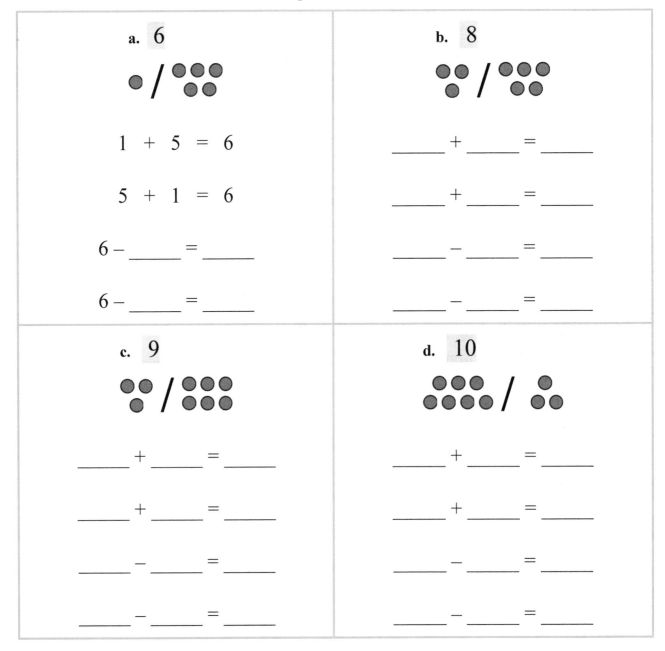

a. 6

● / ●●●
  ●●

$1 + 5 = 6$

$5 + 1 = 6$

$6 - \underline{\hspace{1cm}} = \underline{\hspace{1cm}}$

$6 - \underline{\hspace{1cm}} = \underline{\hspace{1cm}}$

b. 8

●● / ●●●
●   ●●

$\underline{\hspace{1cm}} + \underline{\hspace{1cm}} = \underline{\hspace{1cm}}$

$\underline{\hspace{1cm}} + \underline{\hspace{1cm}} = \underline{\hspace{1cm}}$

$\underline{\hspace{1cm}} - \underline{\hspace{1cm}} = \underline{\hspace{1cm}}$

$\underline{\hspace{1cm}} - \underline{\hspace{1cm}} = \underline{\hspace{1cm}}$

c. 9

●● / ●●●
●   ●●●

$\underline{\hspace{1cm}} + \underline{\hspace{1cm}} = \underline{\hspace{1cm}}$

$\underline{\hspace{1cm}} + \underline{\hspace{1cm}} = \underline{\hspace{1cm}}$

$\underline{\hspace{1cm}} - \underline{\hspace{1cm}} = \underline{\hspace{1cm}}$

$\underline{\hspace{1cm}} - \underline{\hspace{1cm}} = \underline{\hspace{1cm}}$

d. 10

●●● / ●
●●●●   ●●

$\underline{\hspace{1cm}} + \underline{\hspace{1cm}} = \underline{\hspace{1cm}}$

$\underline{\hspace{1cm}} + \underline{\hspace{1cm}} = \underline{\hspace{1cm}}$

$\underline{\hspace{1cm}} - \underline{\hspace{1cm}} = \underline{\hspace{1cm}}$

$\underline{\hspace{1cm}} - \underline{\hspace{1cm}} = \underline{\hspace{1cm}}$

2. Draw circles and write four *different* fact families for which the sum is 7.

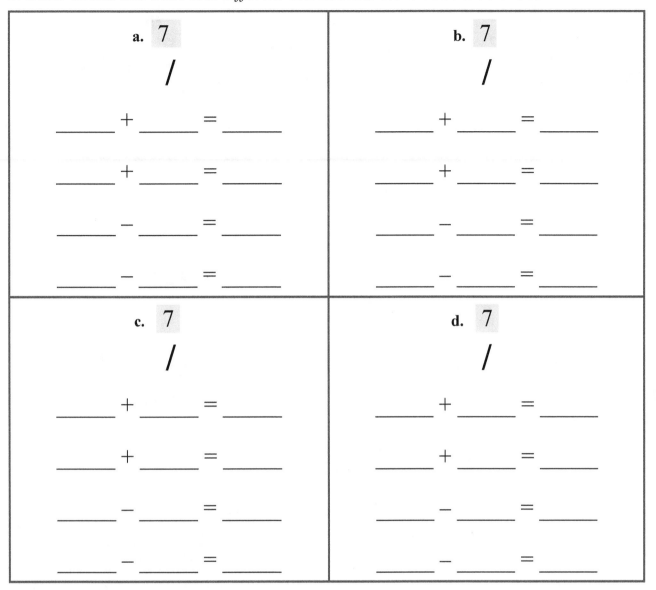

a. 7
/
____ + ____ = ____

____ + ____ = ____

____ − ____ = ____

____ − ____ = ____

b. 7
/
____ + ____ = ____

____ + ____ = ____

____ − ____ = ____

____ − ____ = ____

c. 7
/
____ + ____ = ____

____ + ____ = ____

____ − ____ = ____

____ − ____ = ____

d. 7
/
____ + ____ = ____

____ + ____ = ____

____ − ____ = ____

____ − ____ = ____

3. Ann and Joe solved some math problems that had missing (unknown) numbers.
   Play math teacher. Check their work and correct any mistakes that they made.

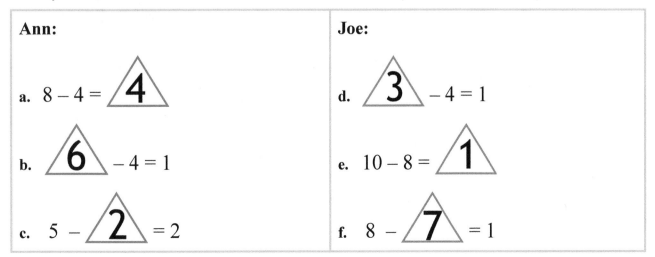

Ann:

a. $8 - 4 = \boxed{4}$

b. $\boxed{6} - 4 = 1$

c. $5 - \boxed{2} = 2$

Joe:

d. $\boxed{3} - 4 = 1$

e. $10 - 8 = \boxed{1}$

f. $8 - \boxed{7} = 1$

4. Make the four facts for the given numbers.

a. Numbers: 5, 3, 2

$$2 + 3 = 5$$

$$3 + 2 = 5$$

$$5 - 3 = 2$$

$$5 - 2 = 3$$

b. Numbers: 9, 4, 5

$$4 + 5 = 9$$

$$\rule{2cm}{0.4pt} + \rule{2cm}{0.4pt} = 9$$

$$9 - 4 = \rule{2cm}{0.4pt}$$

$$9 - \rule{2cm}{0.4pt} = \rule{2cm}{0.4pt}$$

c. Numbers: 4, 0, 4

$$4 + 0 = 4$$

$$\rule{2cm}{0.4pt} + \rule{2cm}{0.4pt} = \rule{2cm}{0.4pt}$$

$$4 - 0 = \rule{2cm}{0.4pt}$$

$$\rule{2cm}{0.4pt} - \rule{2cm}{0.4pt} = \rule{2cm}{0.4pt}$$

d. Numbers: 10, 3, 7

$$\rule{2cm}{0.4pt} + \rule{2cm}{0.4pt} = \rule{2cm}{0.4pt}$$

$$\rule{2cm}{0.4pt} + \rule{2cm}{0.4pt} = \rule{2cm}{0.4pt}$$

$$\rule{2cm}{0.4pt} - \rule{2cm}{0.4pt} = \rule{2cm}{0.4pt}$$

$$\rule{2cm}{0.4pt} - \rule{2cm}{0.4pt} = \rule{2cm}{0.4pt}$$

e. Numbers: 10, \rule{2cm}{0.4pt}, 8

$$\rule{2cm}{0.4pt} + \rule{2cm}{0.4pt} = \rule{2cm}{0.4pt}$$

$$\rule{2cm}{0.4pt} + \rule{2cm}{0.4pt} = \rule{2cm}{0.4pt}$$

$$\rule{2cm}{0.4pt} - \rule{2cm}{0.4pt} = \rule{2cm}{0.4pt}$$

$$\rule{2cm}{0.4pt} - \rule{2cm}{0.4pt} = \rule{2cm}{0.4pt}$$

f. Numbers: 6, 0, \rule{2cm}{0.4pt}

$$\rule{2cm}{0.4pt} + \rule{2cm}{0.4pt} = \rule{2cm}{0.4pt}$$

$$\rule{2cm}{0.4pt} + \rule{2cm}{0.4pt} = \rule{2cm}{0.4pt}$$

$$\rule{2cm}{0.4pt} - \rule{2cm}{0.4pt} = \rule{2cm}{0.4pt}$$

$$\rule{2cm}{0.4pt} - \rule{2cm}{0.4pt} = \rule{2cm}{0.4pt}$$

Sometimes the two addition facts are the same. When that happens, the two subtraction facts will also the same.

For example, with 8, 4, and 4, $4 + 4 = 8$ $8 - 4 = 4$
 we only get one addition fact
 and one subtraction fact. $(4 + 4 = 8)$ $(8 - 4 = 4)$

5. Write the fact families.

a. Numbers: 10, 5, 5

_____ + _____ = _____

_____ + _____ = _____

_____ − _____ = _____

_____ − _____ = _____

b. Numbers: 9, 1, 8

_____ + _____ = _____

_____ + _____ = _____

_____ − _____ = _____

_____ − _____ = _____

c. Numbers: 6, 3, _____

_____ + _____ = _____

_____ + _____ = _____

_____ − _____ = _____

_____ − _____ = _____

d. Numbers: 7, 1, _____

_____ + _____ = _____

_____ + _____ = _____

_____ − _____ = _____

_____ − _____ = _____

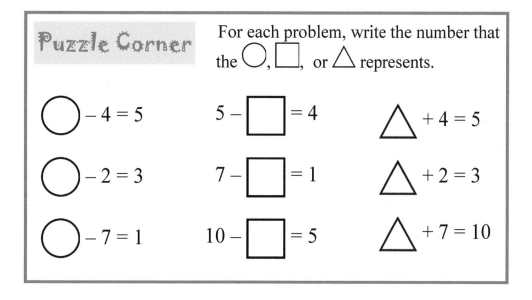

Puzzle Corner   For each problem, write the number that the ◯, ▢, or △ represents.

◯ − 4 = 5   5 − ▢ = 4   △ + 4 = 5

◯ − 2 = 3   7 − ▢ = 1   △ + 2 = 3

◯ − 7 = 1   10 − ▢ = 5   △ + 7 = 10

How Many More?

Henry

| | |

Cindy

Cindy has more marbles.
How many more?

Match three marbles from each box.

Two marbles did not get matched, so Cindy has <u>2 more marbles than</u> Henry.

(Henry has <u>2 fewer marbles than</u> Cindy.)

Henry

| |

Cindy

Cindy has fewer marbles.
How many fewer?

Match two marbles from each box.

Four marbles did not get matched, so Cindy has <u>4 fewer marbles than</u> Henry.

(Henry has <u>4 more marbles than</u> Cindy.)

1. Fill in how many more or how many fewer marbles one child has than another.

Jane

Jim

a. Jane has _____ more than Jim.

Jim has _____ fewer than Jane.

Mark

Mary

b. Mark has _____ more than Mary.

Mary has _____ fewer than Mark.

Ann

Liz

c. Liz has _____ more than Ann.

Ann has _____ fewer than Liz.

Faye

Sam

d. Sam has _____ more than Faye.

Faye has _____ fewer than Sam.

Peter

Frank

e. Frank has _____ more than Peter.

Peter has _____ fewer than Frank.

Susan

Bill

f. Susan has _____ more than Bill.

Bill has _____ fewer than Susan.

2. Now it's your turn to draw. Draw marbles for the child that has none.

[ ] Jane	[◖ ◖] Mark
[◖◖◖◖◖] Jim	[ ] Mary
a. Jane has 3 more than Jim.	**b.** Mary has 4 more than Mark.
[ ] Eric	[◖◖◖◖◖◖] Jack
[◖◖◖◖] Bill	[ ] Jane
c. Eric has 2 fewer than Bill.	**d.** Jane has 5 fewer than Jack.
[ ] Bill	[◖◖◖◖◖◖◖◖] Lucy
[◖◖◖◖◖◖] Greg	[ ] Liz
e. Greg has 1 more than Bill.	**f.** Lucy has 5 more than Liz.
[ ] Ed	[◖◖◖] Ann
[◖◖◖◖◖] Sally	[ ] Mary
g. Sally has 2 fewer than Ed.	**h.** Ann has 4 fewer than Mary.
[ ] Sue	[◖◖◖◖] Jill
[◖◖◖◖◖◖◖] Ben	[ ] Mary
i. Ben has five more than Sue.	**j.** Jill has five fewer than Mary.

3. It is still your turn to draw. You can decide how many marbles the children have.

	Jane		Mark
	Jim		Mary

a. Jane has 5 more than Jim.   **b.** Mary has 2 more than Mark.

	Eric		Jack
	Bill		Jane

c. Eric has 6 fewer than Bill.   **d.** Jane has 7 more than Jack.

	Bill		Lucy
	Greg		Liz

e. Greg has 2 fewer than Bill.   **f.** Lucy has 8 more than Liz.

4. Solve these problems.

a. Ed has five cards, and Jack has seven. How many more cards does Jack have than Ed?

b. John is 8 years old and Jack is 5. How many years older is John?

c. Annie is 10 years old and Beth is 8. How many years younger is Beth?

d. Ruth had 9 dolls, and Tina had 4 dolls. How many fewer did Tina have?

"How Many More" Problems and Differences

How many more emails does Jane have than John?

Draw more emails for John so that the children have the same amount. I drew _____ more emails.

You can write an addition for a "*how many more*" problem.

$2 + \underline{\hspace{1cm}} = 5$

Read: "2 and *how many more* makes five?"

1. Draw more. Read the addition sentences.

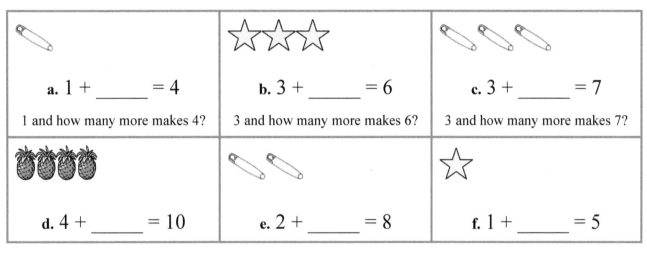

a. $1 + \underline{\hspace{1cm}} = 4$

1 and how many more makes 4?

b. $3 + \underline{\hspace{1cm}} = 6$

3 and how many more makes 6?

c. $3 + \underline{\hspace{1cm}} = 7$

3 and how many more makes 7?

d. $4 + \underline{\hspace{1cm}} = 10$

e. $2 + \underline{\hspace{1cm}} = 8$

f. $1 + \underline{\hspace{1cm}} = 5$

The problem $\underline{\hspace{1cm}} + 2 = 5$ is also read as "2 and *how many more* makes five?"

2. Draw more. Solve.

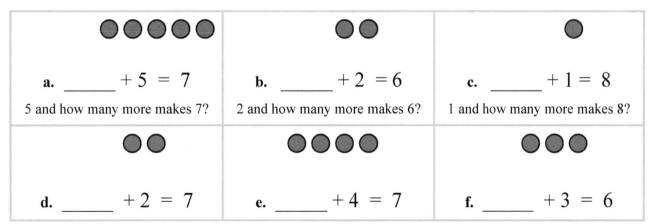

a. $\underline{\hspace{1cm}} + 5 = 7$

5 and how many more makes 7?

b. $\underline{\hspace{1cm}} + 2 = 6$

2 and how many more makes 6?

c. $\underline{\hspace{1cm}} + 1 = 8$

1 and how many more makes 8?

d. $\underline{\hspace{1cm}} + 2 = 7$

e. $\underline{\hspace{1cm}} + 4 = 7$

f. $\underline{\hspace{1cm}} + 3 = 6$

How many more problems are DIFFERENCE problems. The difference of two numbers means <u>how far</u> the two numbers are from each other.

$$0 \quad 1 \quad 2 \quad 3 \quad 4 \quad 5 \quad 6 \quad 7 \quad 8 \quad 9 \quad 10 \quad 11 \quad 12 \quad 13 \quad 14 \quad 15 \quad 16 \quad 17$$

How far is 3 from 7?

How much difference is there between 3 and 7?

3 and how many more make 7?

$3 + \rule{1.5cm}{0.4pt} = 7$

The answer to all these is 4.

How far is 8 from 13?

How much difference is there between 8 and 13?

8 and how many more make 13?

$8 + \rule{1.5cm}{0.4pt} = 13$

The answer to all these is 5.

3. What is the difference between the numbers? Take steps on the number line.

$$0 \quad 1 \quad 2 \quad 3 \quad 4 \quad 5 \quad 6 \quad 7 \quad 8 \quad 9 \quad 10 \quad 11 \quad 12 \quad 13 \quad 14 \quad 15 \quad 16 \quad 17$$

a. from 6 to 10	**b.** from 5 to 8	**c.** from 7 to 11	**d.** from 5 to 5	**e.** from 1 to 10
_____ steps	_____ steps	_____ steps	_____ steps	_____ steps

4. Find the difference between the numbers. "Travel" on the number line!

From	8	4	1	3	6	10	8	9
To	10	10	9	1	5	5	12	15
Difference								

5. Solve the difference between the numbers. Then write an addition. Be careful.

a. from 3 to 5	**b.** from 1 to 5	**c.** from 2 to 7
_____ steps	_____ steps	_____ steps
$3 + \underline{\ 2\ } = 5$	$1 + \rule{1cm}{0.4pt} = 5$	$2 + \rule{1cm}{0.4pt} = 7$

6. Solve the difference between the numbers. Then write an addition. Be careful.

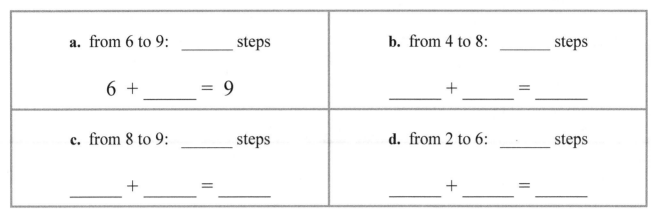

a. from 6 to 9: _____ steps	**b.** from 4 to 8: _____ steps
6 + _____ = 9	_____ + _____ = _____
c. from 8 to 9: _____ steps	**d.** from 2 to 6: _____ steps
_____ + _____ = _____	_____ + _____ = _____

7. Who has more marbles? How many more? Write a "how many more" addition.
   You can also draw.

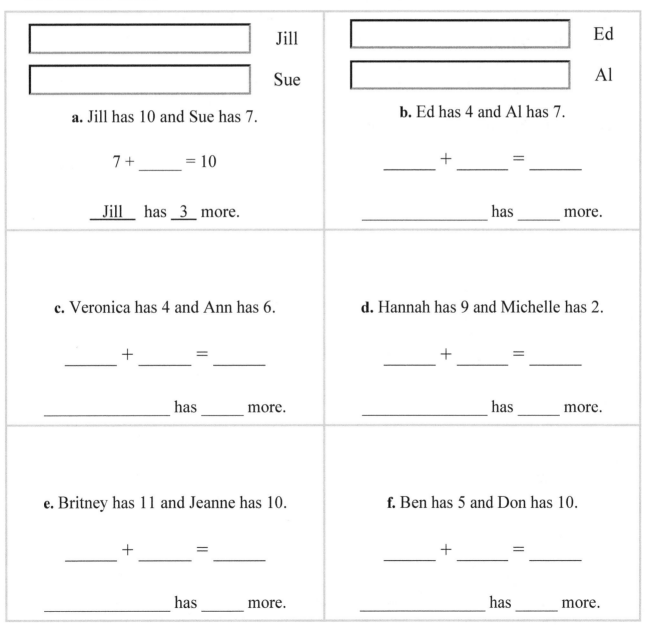

Jill	Ed
Sue	Al
a. Jill has 10 and Sue has 7.	**b.** Ed has 4 and Al has 7.
7 + _____ = 10	_____ + _____ = _____
__Jill__ has _3_ more.	_____ has _____ more.
c. Veronica has 4 and Ann has 6.	**d.** Hannah has 9 and Michelle has 2.
_____ + _____ = _____	_____ + _____ = _____
_____ has _____ more.	_____ has _____ more.
e. Britney has 11 and Jeanne has 10.	**f.** Ben has 5 and Don has 10.
_____ + _____ = _____	_____ + _____ = _____
_____ has _____ more.	_____ has _____ more.

8. Solve the problems. Think carefully: Is it asking for the <u>total</u>?
   OR is it asking, "How many <u>more</u>?"

a. There are two cassette tapes on the table, and eight on the shelf.

How many tapes are there together?

How many more tapes are on the shelf?

b. There are five birds in the apple tree, and
there are five more birds in the oak tree.
Four more flew into the oak tree.

How many birds are now in the oak tree?

How many more birds are in the oak tree than in the apple tree?

c. Brenda has 2 toy cars, Jason has 9, and Joe has 10.

How many more does Joe have than Jason?

How many more does Jason have than Brenda?

"How Many More" Problems and Subtraction

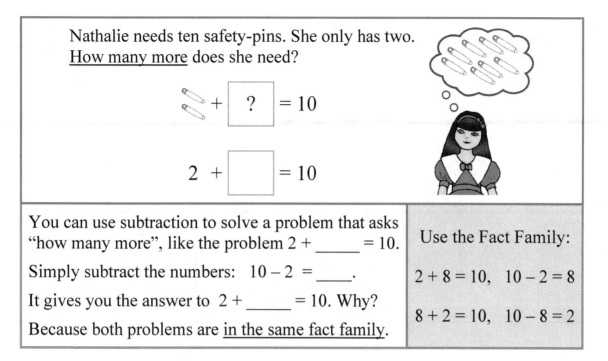

Nathalie needs ten safety-pins. She only has two.
<u>How many more</u> does she need?

$$\text{[pin]} + \boxed{?} = 10$$

$$2 + \boxed{} = 10$$

You can use subtraction to solve a problem that asks "how many more", like the problem $2 + \rule{1cm}{0.15mm} = 10$.

Simply subtract the numbers: $10 - 2 = \rule{1cm}{0.15mm}$.

It gives you the answer to $2 + \rule{1cm}{0.15mm} = 10$. Why?

Because both problems are <u>in the same fact family</u>.

Use the Fact Family:

$2 + 8 = 10, \quad 10 - 2 = 8$

$8 + 2 = 10, \quad 10 - 8 = 2$

1. Fill in. The "how many more" problem has the same answer as the subtraction problem!

[carrots] $+ \boxed{}$	[pickles] $+ \boxed{}$	[carrots] $+ \boxed{}$
a. 5 $+ \rule{1cm}{0.15mm}$ $= 7$	**b.** 3 $+ \rule{1cm}{0.15mm}$ $= 8$	**c.** 4 $+ \rule{1cm}{0.15mm}$ $= 9$
$7 - 5 = \rule{1cm}{0.15mm}$	$8 - 3 = \rule{1cm}{0.15mm}$	$9 - 4 = \rule{1cm}{0.15mm}$
[pickles] $+ \boxed{}$	[pineapples] $+ \boxed{}$	[circles] $+ \boxed{}$
d. 5 $+ \rule{1cm}{0.15mm}$ $= 10$	**e.** 4 $+ \rule{1cm}{0.15mm}$ $= 7$	**f.** 5 $+ \rule{1cm}{0.15mm}$ $= 8$
$10 - 5 = \rule{1cm}{0.15mm}$	$7 - 4 = \rule{1cm}{0.15mm}$	$8 - 5 = \rule{1cm}{0.15mm}$
g. 3 $+ \rule{1cm}{0.15mm}$ $= 10$	**h.** 2 $+ \rule{1cm}{0.15mm}$ $= 9$	**i.** 1 $+ \rule{1cm}{0.15mm}$ $= 7$
$10 - 3 = \rule{1cm}{0.15mm}$	$9 - 2 = \rule{1cm}{0.15mm}$	$7 - 1 = \rule{1cm}{0.15mm}$

2. Solve the subtraction problem *first*. (It is probably easier.) Then copy the answer to the "how many more" problem.

a. $2 +$ ____ $= 8$	**b.** $1 +$ ____ $= 9$	**c.** ____ $+ 3 = 10$	**d.** ____ $+ 3 = 9$
$8 - 2 =$ ____	$9 - 1 =$ ____	$10 - 3 =$ ____	$9 - 3 =$ ____

The "how many more" problem has the same answer as the subtraction problem!

3. Write a subtraction problem, using the same numbers, under each "how many more" problem. Solve the subtraction problem *first*.

a. $1 +$ ____ $= 7$ ____ $-$ ____ $=$ ____	**b.** $2 +$ ____ $= 9$ ____ $-$ ____ $=$ ____	**c.** $1 +$ ____ $= 10$ ____ $-$ ____ $=$ ____
d. ____ $+ 3 = 8$ ____ $-$ ____ $=$ ____	**e.** ____ $+ 2 = 10$ ____ $-$ ____ $=$ ____	**f.** ____ $+ 3 = 9$ ____ $-$ ____ $=$ ____

4. Solve. Think: Do you already know the total? Or is the problem asking for the total? You can also draw a picture to help!

a. Mary ate two carrots. The rabbit ate six carrots. What was the total number of carrots eaten?	**b.** Baby put three blocks in a stack, and another four blocks in another stack. How many blocks did the baby use?
c. There were five lambs in the pen. Two more lambs went into the pen. How many lambs are now in the pen?	**d.** Lisa needs 8 dollars for a stuffed hippo. She has saved 4 dollars. How many more dollars does she need?

5. First write a "how many more" problem for each subtraction problem, using the same numbers. Then solve the easier problem. Copy the answer to the other problem.

a. $8 - 6 =$ ____ ____ $+$ ____ $=$ ____	**b.** $10 - 9 =$ ____ ____ $+$ ____ $=$ ____	**c.** $9 - 7 =$ ____ ____ $+$ ____ $=$ ____
d. $10 - 8 =$ ____ ____ $+$ ____ $=$ ____	**e.** $9 - 8 =$ ____ ____ $+$ ____ $=$ ____	**f.** $7 - 6 =$ ____ ____ $+$ ____ $=$ ____

6. Solve. Think: Do you already know the total? Or is the problem asking for the total? Write an addition or subtraction for each problem. You can also draw a picture to help!

a. Mom needs six cucumbers. She already has three. How many more does she need?	**b.** There were seven ducks on the pond. Three flew away. How many were left?
c. Jane wants to buy a teddy bear for $8. She has saved $6. How much more money does she need?	**d.** A book has ten pages. Jerry has read six pages. How many pages does he have left to read?

7. Play math teacher again. Bill joined Ann and Joe to work some more problems. Check their work and correct any mistakes that they made.

Ann:	Joe:	Bill:
a. $7 - \boxed{1} = 6$	**c.** $\boxed{2} - 4 = 2$	**e.** $\boxed{9} - 4 = 5$
b. $8 - \boxed{3} = 4$	**d.** $9 - \boxed{6} = 3$	**f.** $9 - \boxed{8} = 2$

8. These are the toys that Zach and Mary have.

    a. How many dolls do the children have?

    b. How many teddy bears?

    c. How many other toys?

    Zach and Mary want to make a graph of their toys. To complete the graph, draw one block for each toy. Draw them all the same size and lined up in a column, just like the ones for the dolls.

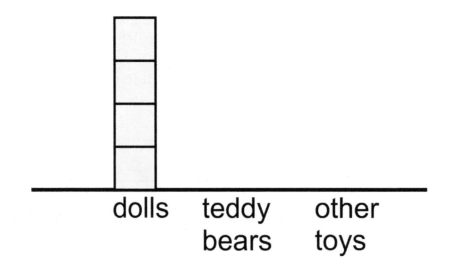

    d. How many more teddy bears do they have than dolls?

    e. How many more teddy bears do they have than other toys?

    f. How many dolls and teddy bears do the children have in all?

Review Chapter 2

1. Write a fact family to match the picture.

_____ + _____ = _____   _____ – _____ = _____

_____ + _____ = _____   _____ – _____ = _____

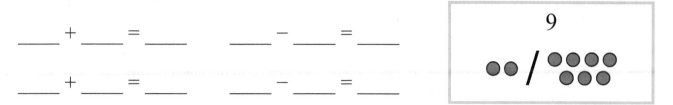

2. **a.** Write a subtraction that matches the addition $6 + 4 = 10$.   ___ – ___ = ___

   b. Write a subtraction that matches the addition $5 +$ ____ $= 9$.   _____ – _____ = _____
   Solve both the addition and the matching subtraction.

3. **a.** There are 8 children playing in the yard. Two are girls.
   How many are boys?

   b. Kay has four marbles. Susan has two more marbles than Kay.
   Draw Kay's and Susan's marbles.

   c. Five sparrows and two robins are feeding on seeds. One more robin flies in.
   How many more sparrows than robins are there now?

4. Find the missing numbers.

a.	b.	c.	d.
$3 +$ ____ $= 4$	$6 - 3 =$ ____	$10 - 0 =$ ____	$8 - 2 =$ ____
$1 +$ ____ $= 9$	$8 - 5 =$ ____	$5 - 3 =$ ____	$7 - 3 =$ ____
$3 +$ ____ $= 10$	$7 - 6 =$ ____	$6 - 6 =$ ____	$10 - 1 =$ ____
$2 +$ ____ $= 7$	$10 - 8 =$ ____	$7 - 4 =$ ____	$9 - 2 =$ ____

Chapter 3: Place Value Within 0-100
Introduction

In this chapter, children learn numbers up to 120. They compare whole numbers within 100, and learn to think of them in terms of tens and ones.

When children count, they basically just learn numbers as some kind of continuum that goes on without end. With simple counting, your child might not catch on to the inherent structure of the number system. Our number system is based on the idea that if you have lots and lots of objects, the efficient way to count and denote them is with *groups* of tens, hundreds, and thousands — not individually.

The crucial point in understanding the concept of place value is therefore that a **certain position represents a group of a specific size**. The digit in each position tells us how many groups of that size there are. For example, in the number 2,381, an adult already knows that the 8 represents eight tens, and not just "8" and that the 3 represents three hundreds, and not just "3". The place of the digit tells us the size of the group, and the digit itself tells how many of that group.

The initial lessons of the book that introduce tens and ones use a **100-bead abacus** extensively. This 100-bead abacus or school abacus simply contains ten beads on ten rods, for a total of 100 beads. It is not the special abacus used by the Chinese or the Japanese. Each bead simply represents one. The 100-bead abacus lets children both "see" the numbers and use their touch while making them.

You will need to purchase this school abacus separately, such as on Amazon, or make your own.

You can browse Amazon's abacus collection at this link:
https://www.amazon.com/s?k=abacus+100+beads&ref=nb_sb_noss_1&tag=mathmammoth-20

Instead of a physical abacus, you can use this online virtual abacus:
https://apps.mathlearningcenter.org/number-rack/

Or, you can make one on your own. This is a fairly easy craft project and you can easily find instructions for it on the Internet (search for example for "DIY abacus").

Besides the abacus, we also use a visual model of blocks where ten of them "snap" together to form a stick. If you already have these so-called base-ten blocks, you can use them along with the visual exercises, if you prefer.

Moreover, we also use number lines and a 100-chart. Number lines help visualize how numbers continue indefinitely and also relate to the concept of measuring. The 100-chart helps the child to be familiar with the numbers below 100 and find patterns in the number system.

While most of the lessons in the book focus on place value, students also practice adding and subtracting multiples of ten and skip-counting. The two lessons at the end of the chapter about tally marks and graphs are real-life applications of two-digit numbers.

The Lessons in Chapter 3

Helpful Resources on the Internet

Use these free online resources to supplement the "bookwork" as you see fit.

You can also access this list of links at https://links.mathmammoth.com/gr1ch3

PLACE VALUE

Online Addition Practice
Practice adding single-digit numbers to multiples of ten with this interactive online exercise.
https://www.mathmammoth.com/practice/place-value#mode=write-number&max-digits=2&question-number=10

Missing Number Addition Practice
Find the missing number in each addition problem in this interactive online exercise.
https://www.mathmammoth.com/practice/place-value#mode=missing-part&max-digits=2&question-number=10

Shark Pool Place Value
Click on the number shown by the ten-stacks and individual blocks.
https://www.ictgames.com/sharkNumbers/mobile/index.html

Base Ten Blocks
Use base ten blocks to help count and show number values.
https://www.roomrecess.com/mobile/BaseTenBlocks/play.html

Name That Number
Match the number on the fruit to the name of the number.
https://www.sheppardsoftware.com/mathgames/earlymath/fruitShootNumbersWords.htm

Fruit Splat Place Value
Click on the fruit that matches the number of tens and ones that are shown. Choose "medium" level.
https://www.sheppardsoftware.com/mathgames/placevalue/fruit_shoot_place_value.htm

Number Bubble Skip Counting
Pop the bubbles so the numbers drop into the correct chests and form a skip-counting pattern.
https://www.abcya.com/games/number_bubble_skip_counting

Lifeguards
Move the boat the correct number of jumps on the number line to save the person.
https://www.ictgames.com/mobilePage/lifeguards/index.html

100-CHART

Count to 99!
Enter the number shown by the colored blocks on a hundred chart.
http://www.thegreatmartinicompany.com/Kids-Math/kids-count-99.html

Give the Dog a Bone
Find the hidden bones on a 100-chart.
https://www.primarygames.co.uk/pg2/dogbone/gamebone.html

Number Charts
Create different kinds of printable number charts.
https://www.homeschoolmath.net/worksheets/number-charts.php

Interactive 100-Chart
Choose a color and create pretty number patterns on this interactive chart.
https://www.abcya.com/games/interactive_100_number_chart

Number Grid Fireworks
Click on the correct square on the number chart to find the hidden fireworks.
https://www.abcya.com/games/100_number_grid

Fill in the Missing Numbers–Customizable Chart
Practice filling in numbers in order, or by types. Set the "End Number" to 120.
https://mrnussbaum.com/the-amazing-number-chart-online

Hundred Chart Game
Answer the questions using the number chart.
https://www.softschools.com/math/hundreds_chart/games/

Interactive Hundred Chart
Color to see skip-counting patterns.
https://www.mathsisfun.com/numbers/number-chart.php

COMPARING

Number Comparison at Mr. Martini's Classroom
Click on the < , > , or = sign to be put in between two numbers.
http://www.thegreatmartinicompany.com/inequalities/number-comparison.html

Counting Caterpillar
Place the leaves in the correct number order for the caterpillar to munch. Choose 0 and 100 as the range.
http://www.ictgames.com/mobilePage/countingCaterpillar/index.html

Order Numbers 1-100 Balloon Pop
Pop the balloons in order from the smallest to the greatest.
https://www.sheppardsoftware.com/mathgames/earlymath/BalloonPopOrder2.htm

Guess the Number
Guess the number to unlock the phone and see the picture! Choose the number range 1-100.
https://www.abcya.com/games/guess_the_number

SKIP-COUNTING

Two-Digit Addition & Subtraction with Mental Math — Online Practice
Choose "2-digit + multiple of 10" and "2-digit - multiple of 10" to practice adding and subtracting whole tens (multiples of tens).
https://www.mathmammoth.com/practice/addition-subtraction-two-digit.php

Add and Subtract with Whole Tens
Practice adding and subtracting with multiples of ten in this interactive online activity.
https://www.mathmammoth.com/practice/addition-subtraction-two-digit#opts=mo10pmo10,mo10mmo10

Skip-Count by 2s – Balloon Rise – Washington Monument
Practice skip-counting by 2s and help the hot-air balloons rise to the top of the Washington Monument.
https://www.free-training-tutorial.com/skip-counting/skip-counting-by-twos-washington-monument.html

Skip-Count by 5s – Balloon Rise – Empire State Building
Practice skip-counting by 5s and help the hot-air balloons rise to the top of the Empire State Building.
https://www.free-training-tutorial.com/skip-counting/skip-counting-by-fives-empire-state.html

Number Line Addition Game
Score goals by clicking on the parachute with the same addition equation as on the number line. Choose level 2.
https://www.turtlediary.com/game/count-up-and-down-by-1-on-number-line.html

Connect the Dots
Connect the dots by counting by twos.
http://www.harcourtschool.com/activity/connect_the_dots/

Froggy Hop
Find 10 more or 1 more than a given number.
https://www.ictgames.com/frog.html

GRAPHS

Bar Graph Sorter
Sort shapes according to shape or color and fill the bar graph.
http://www.shodor.org/interactivate/activities/BarGraphSorter/

Interactive Bar Chart with Questions
Choose a theme and the desired number of intervals for your bar chart. Then, answer the questions.
https://www.topmarks.co.uk/Flash.aspx?f=barchartv2

Tally Chart
Use the tally chart to answer 5 questions.
https://www.softschools.com/math/data_analysis/tally_chart/

Interactive Tally Chart and Bar Graph activity
Click on the children to find out their favorite hobbies. Then make a frequency table and a bar chart from the information.
http://flash.topmarks.co.uk/4771

Counting in Groups of 10

1. Count in groups of TEN. Count ten dots, and circle them.
   Write how many groups of ten you get. Write how many ones are left over.

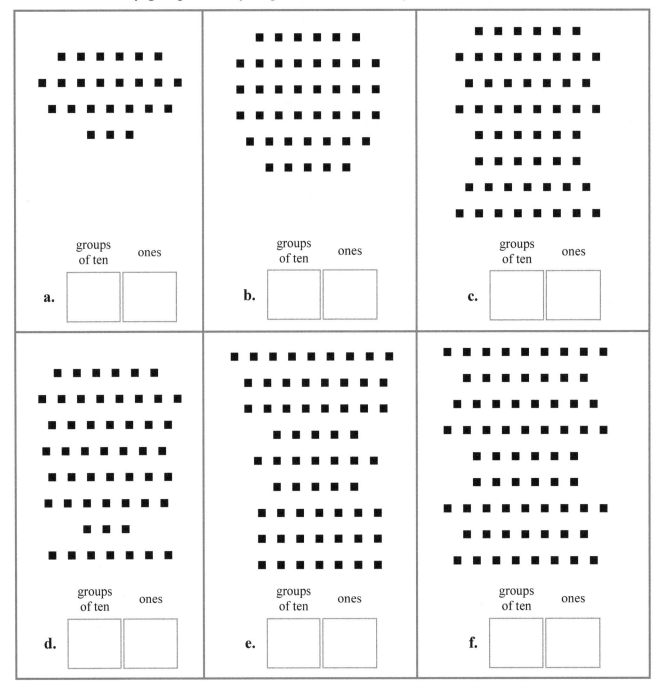

	groups of ten	ones		groups of ten	ones		groups of ten	ones
a.			b.			c.		

	groups of ten	ones		groups of ten	ones		groups of ten	ones
d.			e.			f.		

2. **Counting game 1.** (Optional - to give more practice for making groups of ten.) Put some small items on a table. Ask the child to make groups of ten. Then ask the child to count the groups of ten and the individual items left over, using the form "four tens and six" or "seven tens and one", *etc*. Repeat with different amounts of items, taking turns.

3. **Counting game 2.** <u>You need:</u> Counting items, such as sticks, beads, or beans. Containers, such as small bags or bowls. (If using sticks, then use rubber bands to group them, rather than containers.)

   <u>Before the game:</u> Place 10 of the items in the middle and the rest in a pile on the side.

   <u>Play:</u> At his turn, each player adds one more item to the middle pile on the table, and names the number that is formed. Whenever a whole ten is completed, those ten items are grouped together with a rubber band or by placing them in a small bag or bowl.

   Use ONLY the number words from one to ten when counting in this game. Words like eleven, thirteen, fifty, *etc.* are not allowed. For example, eleven is said as "ten and one", twelve is "ten and two", twenty is "two tens", twenty-five is "two tens and five", *etc.*

   <u>Variation:</u> Each player adds *two* (or more) items to the pile instead of just one.

4. Introduce the 100-bead abacus to the student.
   Make these numbers with the 100-bead abacus.

   a. 6 tens, 5 ones   **e.** 2 tens, 1 one   **i.** 4 tens, 6 ones

   b. 2 tens, 7 ones   **f.** 8 tens, 9 ones   **j.** 6 tens

   c. 7 tens   **g.** 9 tens, 3 ones   **k.** 7 tens, 1 one

   d. 1 ten, 5 ones   **h.** 1 ten, 1 one   **l.** 1 ten, 8 ones

5. Take turns telling each other what number to make on the abacus, such as "7 tens and 9" or "1 ten and 7". **Do not** proceed farther until the student has mastered this! This is crucial.

The names of the numbers with whole tens are:

ten	= ten	four tens	= forty	seven tens	= seventy
two tens	= twenty	five tens	= fifty	eight tens	= eighty
three tens	= thirty	six tens	= sixty	nine tens	= ninety

ten tens = one hundred

---

6. Say the number names from ten to a hundred aloud a few times while also making the numbers on the 100-bead abacus. It almost sounds like a rhyme!

Naming and Writing Numbers

Use the names for the whole tens (twenty, thirty, forty, *etc.*) when naming numbers. Write the number of tens and the number of ones next to each other with numerals.

"6 tens and 7 ones" is sixty-seven, or 67.   "4 tens and 5 ones" is forty-five, or 45.

"8 tens and 1 one" is eighty-one, or 81.   "2 tens and 2 ones" is twenty-two, or 22.

Numbers between 10 and 20 are more difficult. We will look at them in the next lesson.

1. Make these numbers with the 100-bead abacus. Also name them and write them with numbers.

   a. 5 tens 6 ones   *fifty-six*   56

   b. 7 tens 2 ones

   c. 2 tens 1 one

   d. 3 tens 1 one

   e. 4 tens 8 ones

   f. 3 tens 5 ones

   g. 2 tens 3 ones

   h. 8 tens 7 ones

   i. 9 tens 4 ones

   j. 6 tens 6 ones

2. The **Counting game again** - this time with number names from 20 on.

  You need: Counting items, such as sticks, beads, or beans.
  Containers, such as small bags or bowls. (If using sticks, then use rubber bands to group them, rather than containers.)

  Preparation: Place 20 of the items in the middle and the rest in a pile to one side.

  Play: During his turn, each player adds one more item to the middle pile on the table, and names the number that is formed. Whenever a whole ten is completed, those ten items are grouped together with a rubber band or by placing them in a small bag or bowl.

  Variation:
  *Each player adds *two* (or more) items to the pile instead of just one.

3. Take turns telling each other what number to make on the abacus, such as "fifty-six". Do not use the numbers between 10 and 20 yet.

4. Make these numbers with the abacus or with some counting items. Fill in the table.

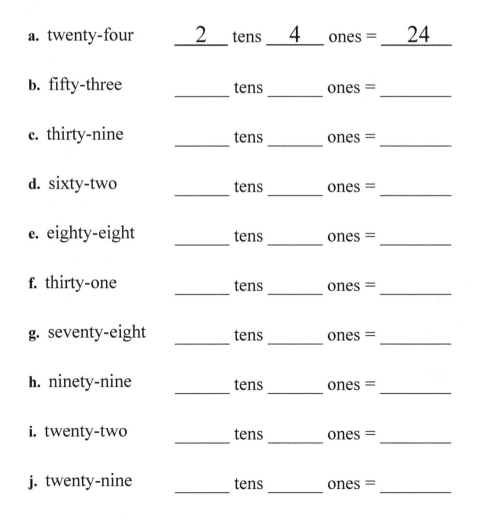

  a. twenty-four   __2__ tens __4__ ones = __24__

  b. fifty-three   _____ tens _____ ones = _____

  c. thirty-nine   _____ tens _____ ones = _____

  d. sixty-two   _____ tens _____ ones = _____

  e. eighty-eight   _____ tens _____ ones = _____

  f. thirty-one   _____ tens _____ ones = _____

  g. seventy-eight   _____ tens _____ ones = _____

  h. ninety-nine   _____ tens _____ ones = _____

  i. twenty-two   _____ tens _____ ones = _____

  j. twenty-nine   _____ tens _____ ones = _____

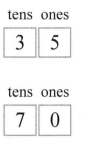

tens	ones	
3	**5**	In the number 35, the "3" means three tens. The "3" is in the tens place. The "5" means five ones. The "5" is in the ones place.

tens	ones	
7	**0**	In the number 70, there are just seven whole tens, and no ones, so write "7" in the tens place and a zero in the ones place.

5. Circle each group of ten dots. Then count the groups of ten and the ones left over. Write and name the number. Notice how quick it is to count the dots this way!

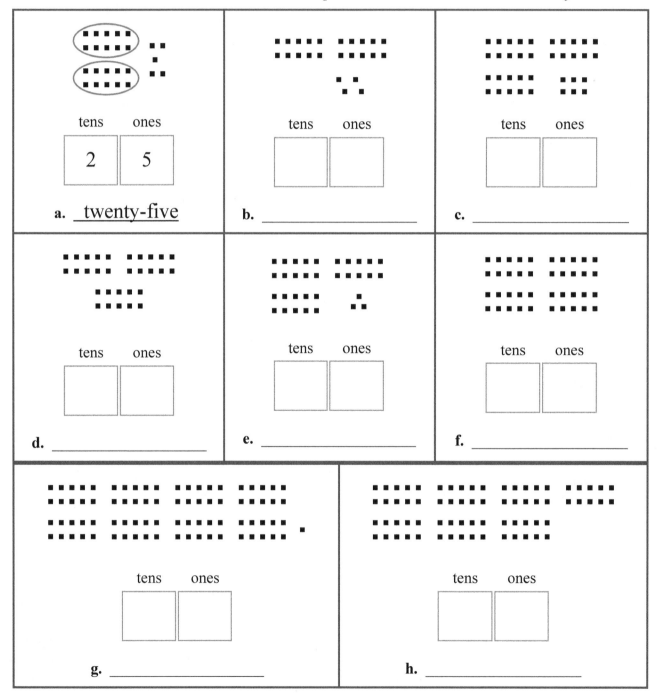

a. __twenty-five__

b. _____

c. _____

d. _____

e. _____

f. _____

g. _____

h. _____

6. Circle each group of ten dots. Then count the groups of ten and the ones left over.
   Write and name the number.

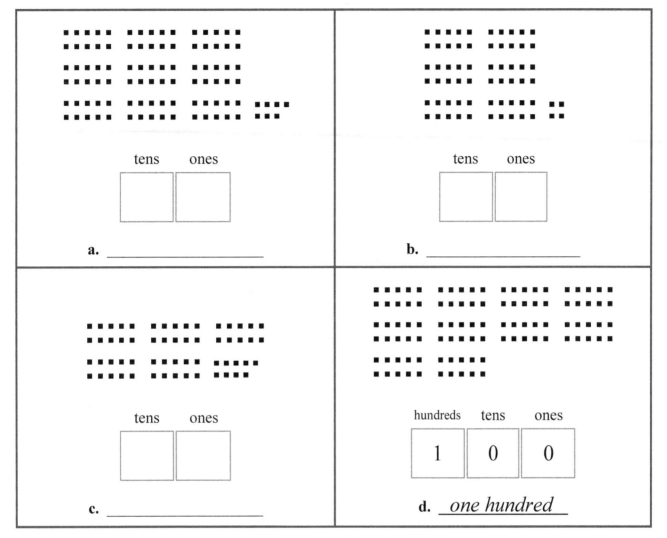

a. _____

tens   ones

b. _____

tens   ones

c. _____

tens   ones

hundreds   tens   ones

| 1 | 0 | 0 |

d. *one hundred*

7. Now draw the dots yourself. Make groups of ten like you did before.

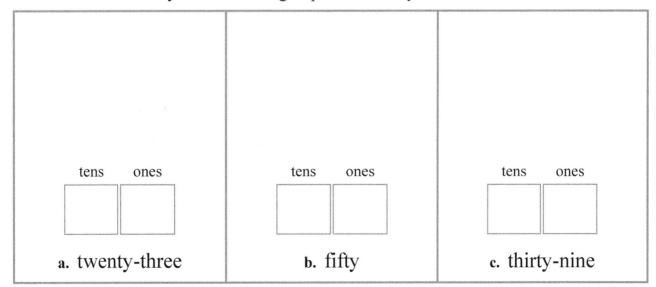

tens   ones

a. twenty-three

tens   ones

b. fifty

tens   ones

c. thirty-nine

The "Teen" Numbers

The names of the numbers between 10 and 20 are:

1 ten 1 = 11 =	eleven	1 ten 6 = 16 = sixteen
1 ten 2 = 12 =	twelve	1 ten 7 = 17 = seventeen
1 ten 3 = 13 =	thirteen	1 ten 8 = 18 = eighteen
1 ten 4 = 14 =	fourteen	1 ten 9 = 19 = nineteen
1 ten 5 = 15 =	fifteen	2 tens = 20 = twenty

The word "teen" actually comes from "ten". So, seventeen is actually "seven-ten", or one ten and seven.

1. Write the numbers on this number line under the tick marks. Say the numbers aloud.

9   10

2. Name and write the numbers.

a. 1 ten 1 one _____ _____

b. 1 ten 7 ones _____ _____

c. 1 ten 2 ones _____ _____

d. 1 ten 6 ones _____ _____

e. 1 ten 3 ones _____ _____

f. 1 ten 8 ones _____ _____

g. 1 ten 4 ones _____ _____

h. 1 ten 5 ones _____ _____

3. Fill in the missing items.

a. 10 + 8 = __18__ _____eighteen_____

b. 10 + 1 = _____ _____

c. 10 + _____ = 14 _____

d. 10 + _____ = 19 _____

e. _____ + 3 = _____ _____

f. 10 + 2 = _____ _____

g. _____ + _____ = 17 _____

h. _____ + _____ = 15 _____

4. This is a number chart from 1 to 100.

a. Find the whole tens on the chart (ten, twenty, thirty, and so on). Color their squares yellow.

b. Find the "teen" numbers on the chart (from thirteen to nineteen). Color their squares pink.

c. The "sixties" row is colored, ending in 70. Find the "fourties" row and color it light green. Color the "nineties" row light brown.

d. Find the column that has all of the numbers that end in "5", such as 5, 15, 25, and so on. Color them light blue.

1	2	3	4	5	6	7	8	9	10
11	12	13	14	15	16	17	18	19	20
21	22	23	24	25	26	27	28	29	30
31	32	33	34	35	36	37	38	39	40
41	42	43	44	45	46	47	48	49	50
51	52	53	54	55	56	57	58	59	60
61	62	63	64	65	66	67	68	69	70
71	72	73	74	75	76	77	78	79	80
81	82	83	84	85	86	87	88	89	90
91	92	93	94	95	96	97	98	99	100

5. Choose one (or both) of the following games to practice the teen numbers some more.

Practice "Teens" board game

You need: A board game where you move a piece along a path. Number cards 1 to 9 plus one number card with 10, from a standard deck of playing cards or UNO cards, etc.

Before the game: Place the card with 10 right-side-up in the middle. Place the deck next to it, face down.

Play: To start his turn, a player draws a card from the deck, placing it next to the 10-card. Then he has to *name* the number formed by the 10-card and the card he drew. For example, if 7 is drawn, the player has to say "seventeen". If he names the number correctly, the player can move his piece the number of steps that was on the card he drew (not including the ten). Otherwise, follow the rules of the board game you are using.

Teens out!

You need: A standard deck of playing cards. Each face card (jack, queen, king) counts as a 10.

Before the game: Deal each person 7 cards. Place the remaining cards (the deck) face down in the middle.

Play: During his turn, the player takes one card from the deck. Then he discards from his hand any two cards that when added together are more than 10 but less than 20, *and* has to name the number when putting the cards down. For example, 10 and 5 make fifteen and can be discarded. Eight and five make thirteen, and can be discarded. Remember, any face card counts as a 10. A player may discard several pairs of cards in one turn. If the player fails to name the number correctly when discarding, he has to keep the cards in his hand. The game ends when someone is able to discard all of his cards from his hand.

Variation: Instead of discarding the cards, the player places a "teen" pair of cards face up in front of him as a "book". The game is played till all the cards are used from the deck. If a person does not have any cards left in his hand, then at his turn, he simply draws one card from the deck and the turn passes to the next person. In the end, whoever has the most "books" wins.

Building Numbers 11-40

1. Fill in the table. Think of the " + " sign as "and": 10 + 3 means 10 *and* 3.

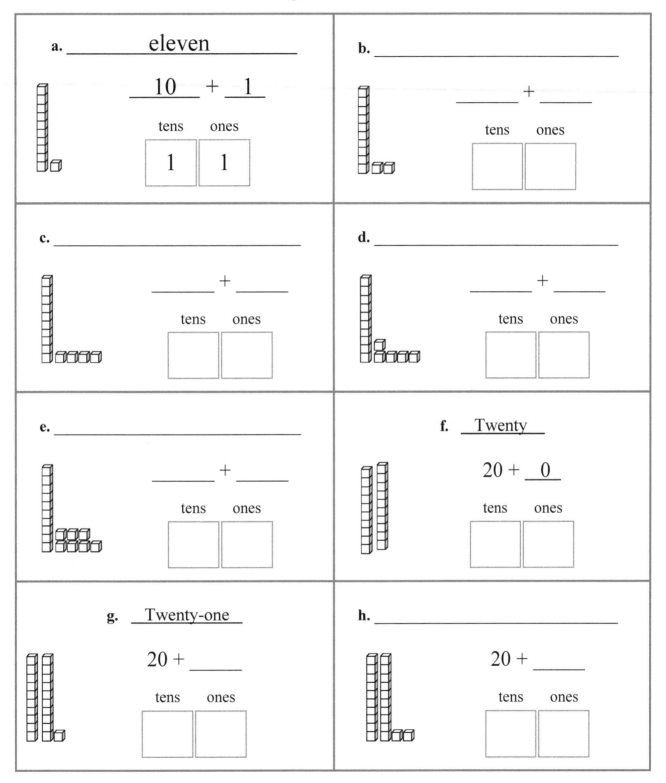

a. _____ eleven _____

__10__ + __1__

tens	ones
1	1

b. _____

_____ + _____

tens	ones

c. _____

_____ + _____

tens	ones

d. _____

_____ + _____

tens	ones

e. _____

_____ + _____

tens	ones

f. __Twenty__

20 + __0__

tens	ones

g. __Twenty-one__

20 + _____

tens	ones

h. _____

20 + _____

tens	ones

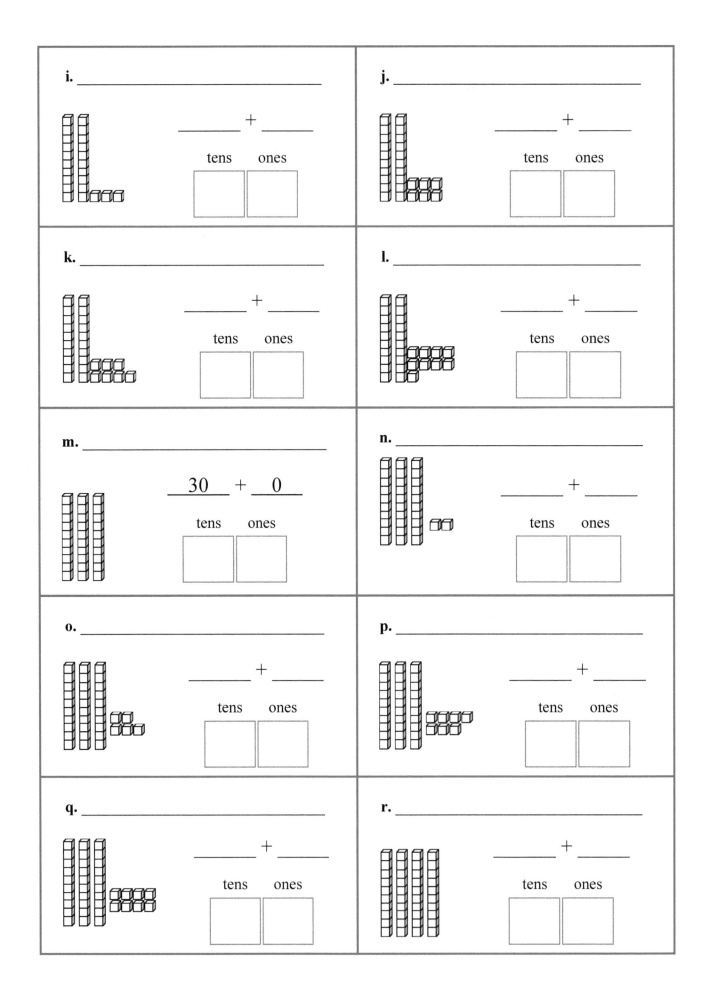

i. _____

___ + ___

tens   ones

j. _____

___ + ___

tens   ones

k. _____

___ + ___

tens   ones

l. _____

___ + ___

tens   ones

m. _____

__30__ + __0__

tens   ones

n. _____

___ + ___

tens   ones

o. _____

___ + ___

tens   ones

p. _____

___ + ___

tens   ones

q. _____

___ + ___

tens   ones

r. _____

___ + ___

tens   ones

Building Numbers 41-100

1. Fill in the table.

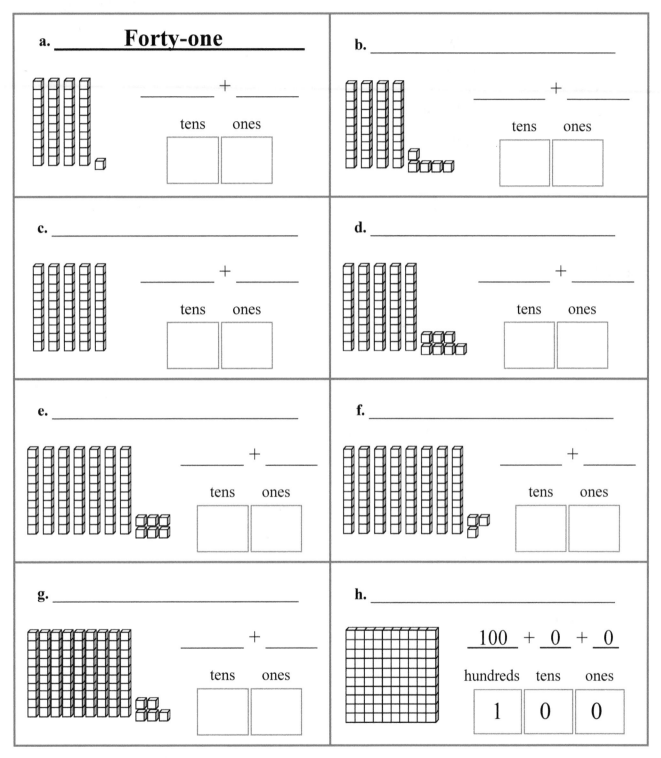

a. ___**Forty-one**___

_____ + _____

tens   ones

b. _____

_____ + _____

tens   ones

c. _____

_____ + _____

tens   ones

d. _____

_____ + _____

tens   ones

e. _____

_____ + _____

tens   ones

f. _____

_____ + _____

tens   ones

g. _____

_____ + _____

tens   ones

h. _____

_100_ + _0_ + _0_

hundreds   tens   ones

| 1 | 0 | 0 |

2. Match the number to its name.

46	forty-two	99	seventy-six
64	forty-six	81	eighty-one
55	fifty-five	90	ninety-one
70	fifty-seven	91	eighty-three
69	fifty-nine	79	ninety
59	sixty-four	76	seventy-nine
42	seventy	100	ninety-nine
57	sixty-nine	83	one hundred

3. Break the numbers into tens and ones.

a. 73 = __70__ + __3__   c. 45 = _____ + _____   e. 98 = _____ + _____

b. 91 = _____ + _____   d. 83 = _____ + _____   f. 64 = _____ + _____

4. Do the same thing the other way around! Add.

a.	b.	c.	d.
60 + 7 = _____	4 + 50 = _____	6 + 80 = _____	90 + 9 = _____
80 + 0 = _____	8 + 80 = _____	0 + 40 = _____	1 + 60 = _____

5. Fill in the missing numbers on the number lines.

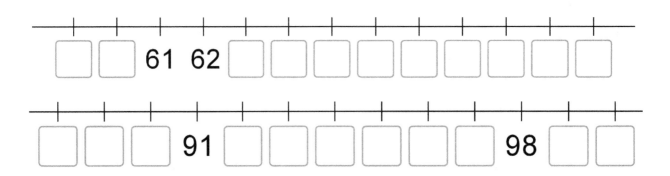

124

A 100-Chart

1. Fill in the numbers missing from this 100-chart.

1		3		5				9	10
11	12							19	
21		23			26				
						38			
41			44						50
	53			56			59		
61			64		67				
	73			77		79			
81		84							
						97	98		100

2. Find these numbers on the chart: 26, 58, 34, 91, 70, and 45. You can color the squares.

3. Practice finding numbers on the chart with your friend or teacher. Take turns giving each other a number to find. Or, play this online game:

   Give the Dog a Bone—Find the hidden bones on a 100-chart.
   http://www.primarygames.co.uk/pg2/dogbone/gamebone.html

4. Write the number that is... (You can also make a game out of this.)

a.	b.	c.
one more than 98 _____	two more than 79 _____	one less than 86 _____
one more than 27 _____	two more than 38 _____	one less than 70 _____

5. Look at the row that starts with 31.
   How are those numbers alike?

6. Look at the column that starts with 5.
   How are those numbers alike?

7. For each highlighted blue number,
   find the number that is exactly
   ten more, and color it.
   Hint: It is exactly under it, in the next row.

8. For each underlined number,
   find the number that is exactly
   ten less, and underline it.
   Hint: It is exactly above it, in the previous row.

1	2	3	4	5	6	7	8	9	10
11	12	**13**	14	15	16	17	18	19	20
21	22	23	24	25	26	**27**	28	29	30
31	32	33	34	35	36	37	38	39	**40**
41	42	43	44	45	46	47	48	49	50
51	52	53	54	55	56	**57**	58	59	60
61	62	63	64	65	66	67	68	69	70
71	72	73	74	75	76	77	78	79	80
81	82	83	**84**	85	86	87	88	89	90
91	92	93	94	95	96	97	98	**99**	100

> What number is 10 more than 49? In other words, what is 49 + 10?
> Simply look at the tens number: **4**9, and add one to it. You get 59.

> What number is 10 less than 72? In other words, what is 72 − 10?
> Look at the tens number: **7**2, and take away one from it. You get 62.

9. Write the number that is...

a.	**b.**	**c.**
ten more than 18 _____	ten more than 60 _____	ten less than 24 _____
ten more than 35 _____	ten more than 59 _____	ten less than 37 _____

10. Fill in the numbers missing from these number lines.

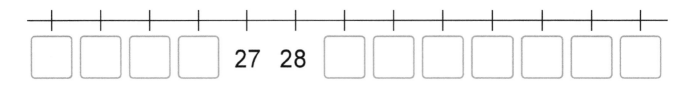

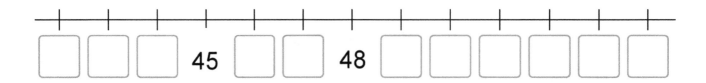

126

Add and Subtract Whole Tens

1. Add and subtract with tens. You can also solve these with the abacus.

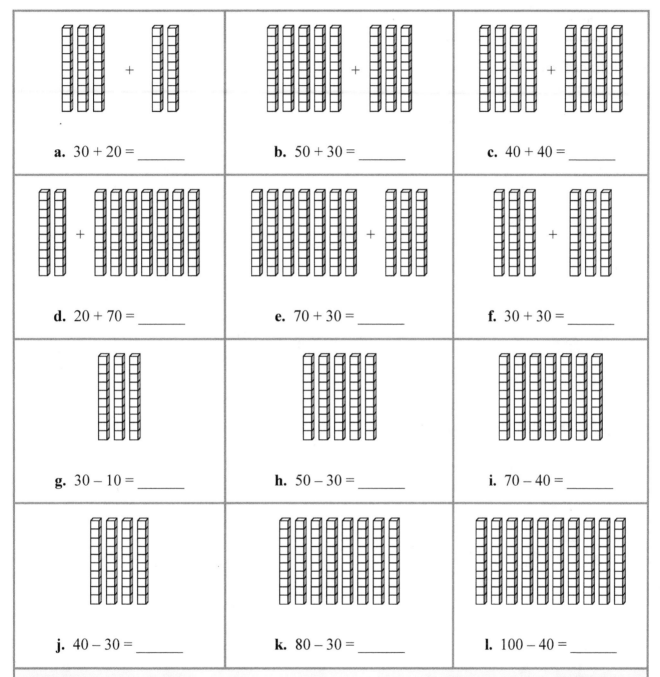

a. 30 + 20 = _____

b. 50 + 30 = _____

c. 40 + 40 = _____

d. 20 + 70 = _____

e. 70 + 30 = _____

f. 30 + 30 = _____

g. 30 – 10 = _____

h. 50 – 30 = _____

i. 70 – 40 = _____

j. 40 – 30 = _____

k. 80 – 30 = _____

l. 100 – 40 = _____

What do you notice?

Adding **60 + 30** means **6 tens + 3 tens** = 9 tens.
So I can simply add 6 + 3 = 9, and just remember the answer will be 9 tens, or 90.

Subtracting **80 – 70** means **8 tens – 7 tens** = 1 ten.
I can simply subtract 8 – 7 = 1, and just remember the answer will be 1 ten, or 10.

2. Now, try these *without* using the abacus or other help.

a. $10 + 50 =$ _____	**b.** $60 + 30 =$ _____	**c.** $70 - 10 =$ _____
$60 + 20 =$ _____	$20 + 50 =$ _____	$90 - 40 =$ _____

3. Subtract from 100. Use the abacus if you need to.

a. $100 - 50 =$ _____	**b.** $100 - 30 =$ _____	**c.** $100 - 90 =$ _____
$100 - 20 =$ _____	$100 - 70 =$ _____	$100 - 40 =$ _____

4. Fill in the missing numbers.

a. $60 + \boxed{} = 90$	**b.** $80 - \boxed{} = 30$	**c.** $\boxed{} - 10 = 10$
$20 + \boxed{} = 100$	$80 - \boxed{} = 40$	$\boxed{} - 20 = 20$

5. Add the same number several times.

a. $10 + 10 + 10 + 10 + 10 =$ _____   **b.** $20 + 20 + 20 + 20 + 20 =$ _____

6. Add and subtract whole tens.

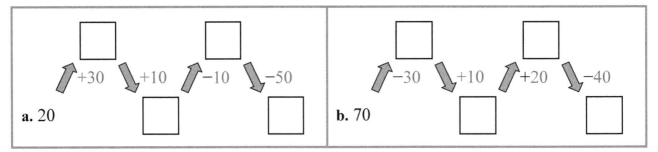

a. 20

b. 70

Puzzle Corner   *Quick additions!* In these sums, first add any numbers that make ten. (We can do that because we can add in any order.)

$10 + 2 + 20 + 4 + 5 + 1 + 8 + 9 + 9 + 20 + 5 + 6 =$ _____

$8 + 6 + 7 + 9 + 5 + 30 + 4 + 3 + 5 + 2 + 1 + 20 + 1 =$ _____

Practicing with Numbers

1. Fill in.

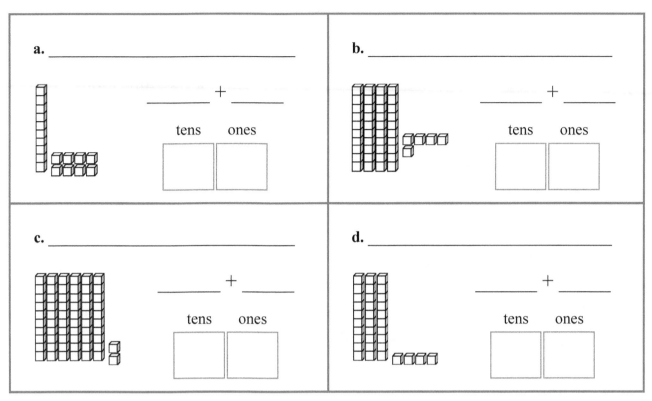

a. _____

_____ + _____
tens   ones

b. _____

_____ + _____
tens   ones

c. _____

_____ + _____
tens   ones

d. _____

_____ + _____
tens   ones

2. Add ten by drawing another column of ten. Subtract ten by crossing out a column.

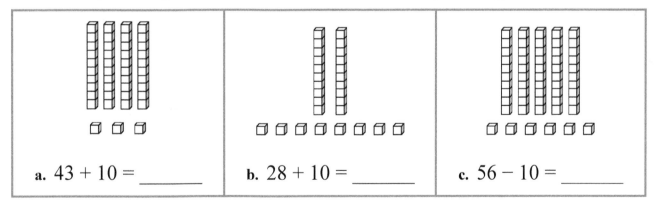

a. 43 + 10 = _____

b. 28 + 10 = _____

c. 56 − 10 = _____

3. *Explain* to your teacher how you add a ten to a number, or subtract a ten.

a. 61 + 10 = _____	b. 37 + 10 = _____	c. 89 + 10 = _____
61 − 10 = _____	37 − 10 = _____	89 − 10 = _____

4. Match the number to its name.

16	twelve	90	eleven
60	fourteen	29	nineteen
23	sixteen	11	ninety-three
32	thirty-two	19	ninety-one
76	sixty	91	twenty-nine
14	twenty-three	39	thirty-nine
12	seventy-six	93	ninety

5. Write each number as (tens) + (ones).

a. 91 = _____ + _____   b. 79 = _____ + _____   c. 58 = _____ + _____

6. Add.

a. 20 + 6 = _____   b. 60 + 7 = _____   c. 6 + 50 = _____

80 + 2 = _____   10 + 1 = _____   5 + 30 = _____

7. Fill in.

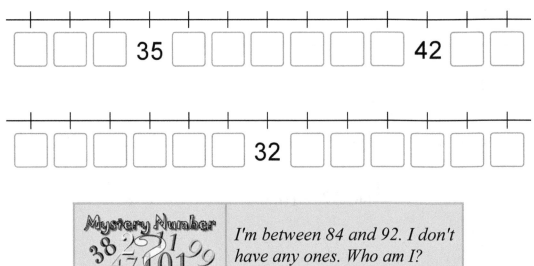

I'm between 84 and 92. I don't have any ones. Who am I?

130

Which Number Is Greater?

1. The pictures have ten-sticks and one-dots. Fill in the missing parts.
   Then circle the number that is **more**.

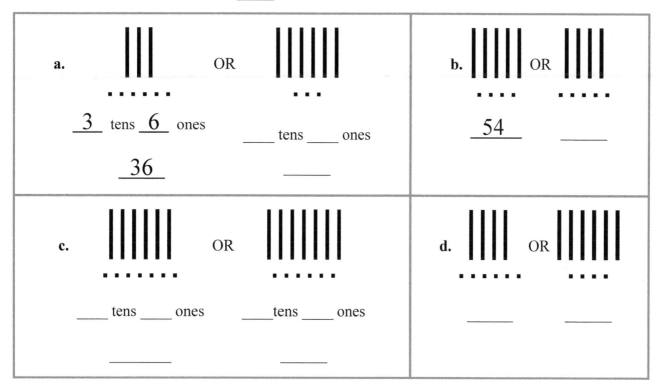

Study the above pictures. Do we first check how many TENS the numbers have or how many ONES the numbers have?

- Check first how many _____ the numbers have.

- If the numbers have the same amount of _____, then compare the _____.

For example:

- 92 has more TENS than 89, so 92 is greater than 89.

- 62 has the same amount of TENS as 66, but it has less ONES than 66.
  Therefore 62 is less than 66.

The symbol > means "greater than", and < means "less than". Think of it as a hungry alligator's mouth! The open end (the open "mouth") of the symbol ALWAYS points to the **bigger** number—the alligator wants to eat as much as it can!

For example: 3 < 5   14 > 3   60 > 50   48 < 99   7 < 17

2. Write < or > between the numbers. You can draw ten-sticks and one-dots to help.

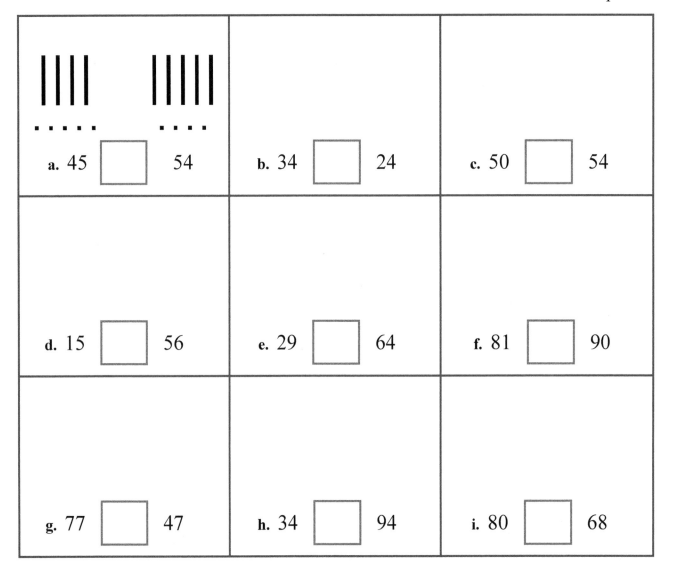

a. 45 ☐ 54

b. 34 ☐ 24

c. 50 ☐ 54

d. 15 ☐ 56

e. 29 ☐ 64

f. 81 ☐ 90

g. 77 ☐ 47

h. 34 ☐ 94

i. 80 ☐ 68

3. Write < or > between the numbers to compare them. Number line can help.

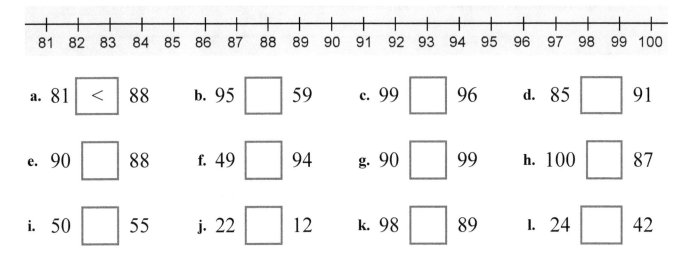

a. 81 < 88

b. 95 ☐ 59

c. 99 ☐ 96

d. 85 ☐ 91

e. 90 ☐ 88

f. 49 ☐ 94

g. 90 ☐ 99

h. 100 ☐ 87

i. 50 ☐ 55

j. 22 ☐ 12

k. 98 ☐ 89

l. 24 ☐ 42

4. Write <, > or = .

a. $80 + 2$ $\boxed{<}$ $80 + 7$ b. $10 + 7$ $\boxed{\phantom{<}}$ $70 + 5$

c. $60 + 5$ $\boxed{\phantom{<}}$ $40 + 5$ d. $50 + 7$ $\boxed{\phantom{<}}$ $70 + 3$

e. $6 + 60$ $\boxed{\phantom{<}}$ $6 + 80$ f. $5 + 40$ $\boxed{\phantom{<}}$ $40 + 5$

g. $40 + 5$ $\boxed{\phantom{<}}$ $40 + 4$ h. $70 + 5$ $\boxed{\phantom{<}}$ $5 + 70$

5. Put the numbers in order from the smallest to the largest.

a. 68, 67, 46	b. 33, 53, 37
_____ < _____ < _____	_____ < _____ < _____
c. 48, 46, 18	d. 87, 78, 80
_____ < _____ < _____	_____ < _____ < _____
e. 48, 84, 46, 50	f. 98, 87, 67, 76
_____ < _____ < _____ < _____	_____ < _____ < _____ < _____

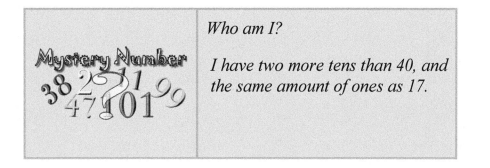

Who am I?

I have two more tens than 40, and the same amount of ones as 17.

133

Numbers Past 100

Counting past 100 is very easy. You just start over, but instead of 1, 2, 3, 4, and so on, you count, "one hundred and one", "one hundred and two", "one hundred and three", "one hundred and four", and so on.

Count aloud with your teacher from one hundred to one hundred twenty.
Look at how these numbers are written:

100, 101, 102, 103, 104, 105, 106, 107, 108, 109, 110,
111, 112, 113, 114, 115, 116, 117, 118, 119, 120.

1. Fill in the missing numbers. This chart starts at one hundred one (101) and ends at 130.

101	102	103	104						110
111	112	113					118	119	120
121	122	123	124	125	126	127	128	129	130

2. Fill in the missing numbers. This chart starts at 81 and ends at 110.

81	82			85	86		88		
91			94			97			
		103			106			109	110

3. Write the number that is... (You can use the above charts to help.)

a.	b.	c.
one more than 109 _____ one more than 113 _____	two more than 116 _____ two more than 99 _____	ten more than 100 _____ ten more than 102 _____

d.	e.	f.
one less than 108 _____ one less than 120 _____	two less than 110 _____ two less than 117 _____	ten less than 120 _____ ten less than 107 _____

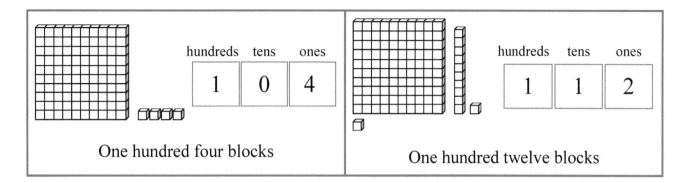

One hundred four blocks

hundreds	tens	ones
1	0	4

One hundred twelve blocks

hundreds	tens	ones
1	1	2

4. Fill in the table.

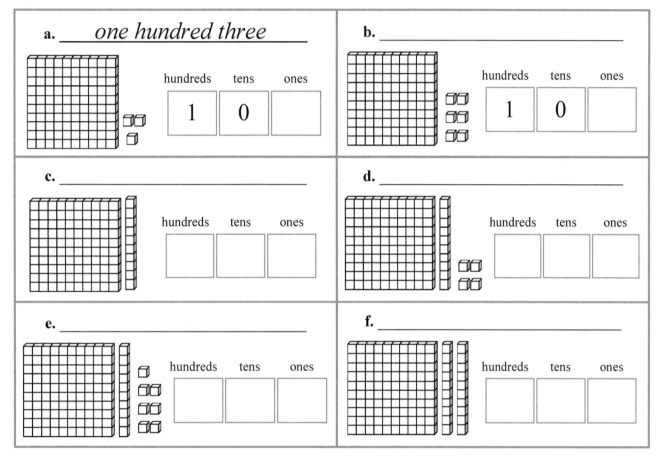

a. *one hundred three*

hundreds	tens	ones
1	0	

b. _____

hundreds	tens	ones
1	0	

c. _____

hundreds	tens	ones

d. _____

hundreds	tens	ones

e. _____

hundreds	tens	ones

f. _____

hundreds	tens	ones

5. Fill in the missing numbers. You can also say this aloud with your teacher.

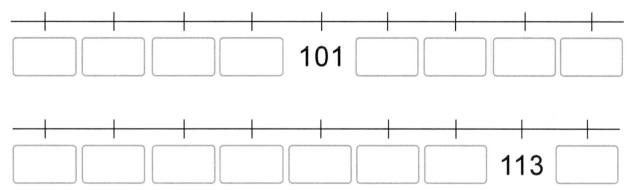

101

113

More Practice with Numbers

1. Fill in the missing numbers on the chart.

31	32		34					39	
		43	44	45	46			49	
51	52	53				57			60
			64	65	66		68		
				76	77	78	79		
		83							
	92	93	94	95	96				
101				105					110
					116	117			120

2. Use the chart above to help, and write the number that is...

a.	b.	c.
ten more than 67 _____	ten more than 78 _____	ten less than 55 _____
ten more than 87 _____	ten more than 101 _____	ten less than 70 _____
ten more than 45 _____	ten more than 99 _____	ten less than 118 _____

3. Draw another column of ten in each picture. That's adding 10! Then write the addition sentence. You can also use an abacus.

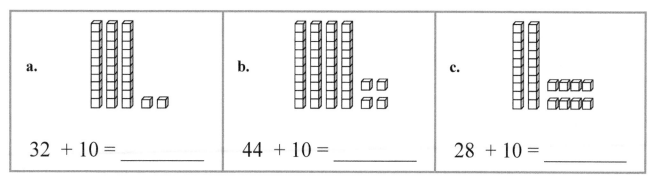

a. 32 + 10 = _____

b. 44 + 10 = _____

c. 28 + 10 = _____

4. Add 10. Use the 100-bead abacus to help.

a. 51 + 10 = _____	b. 16 + 10 = _____	c. 67 + 10 = _____
d. 74 + 10 = _____	e. 65 + 10 = _____	f. 39 + 10 = _____

5. Add the same number several times. Write the addition sentence.

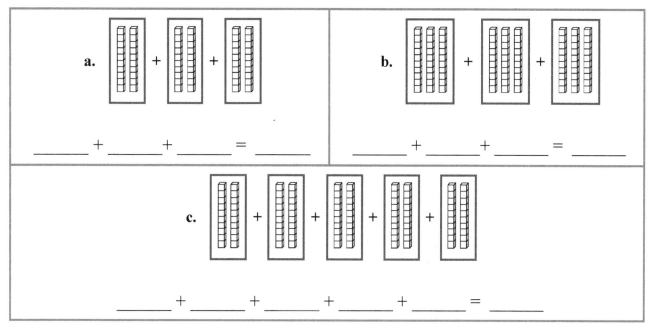

a. _____ + _____ + _____ = _____

b. _____ + _____ + _____ = _____

c. _____ + _____ + _____ + _____ + _____ = _____

6. Count by ones to fill in the missing numbers.

a. 88, 89, _____, _____

b. _____, 16, _____, 18

c. _____, 62, 63, _____

d. _____, 108, _____, 110

Skip-Counting Practice

1. Count by twos. You can also say this aloud with your teacher.

   a. 0, 2, 4, _____, _____, _____, _____, _____, _____, _____, _____

   b. 50, 52, _____, _____, _____, _____, _____, _____, _____, _____

   c. 1, 3, 5, _____, _____, _____, _____, _____, _____, _____, _____

   d. _____, _____, _____, _____, 35, 37, _____, _____, _____, _____

2. Fill in the chart.

   Then color every fourth number on the number chart starting at 4. It should create a pretty pattern.

1	2	3	4	5	6	7	8		
									20
	22	23	24						30
31	32	33	34	35	36	37	38		
41							48	49	50
51	52	53				57			
61			64	65			68		
			75	76	77				
	82	83				87	88	89	
91	92	93			96	97	98	99	100

3. Count by fours. You can also say this aloud with your teacher.

   a. 0, 4, _____, _____, _____, _____, _____, _____, _____, _____

   b. 52, 56, _____, _____, _____, _____, _____, _____, _____, _____

   c. 1, 5, _____, _____, _____, _____, _____, _____, _____, _____

4. Color every third number on the number chart starting at 3. It should create a pretty pattern.

1	2	3	4	5	6	7	8	9	10
11	12	13	14	15	16	17	18	19	20
21	22	23	24	25	26	27	28	29	30
31	32	33	34	35	36	37	38	39	40
41	42	43	44	45	46	47	48	49	50
51	52	53	54	55	56	57	58	59	60
61	62	63	64	65	66	67	68	69	70
71	72	73	74	75	76	77	78	79	80
81	82	83	84	85	86	87	88	89	90
91	92	93	94	95	96	97	98	99	100

5. Count by tens. Use the chart to help. Notice the patterns!

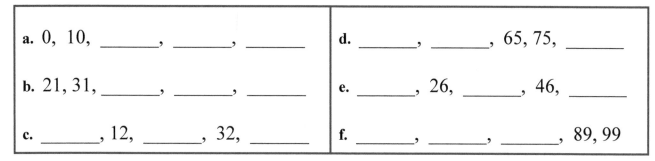

a. 0, 10, _____, _____, _____

b. 21, 31, _____, _____, _____

c. _____, 12, _____, 32, _____

d. _____, _____, 65, 75, _____

e. _____, 26, _____, 46, _____

f. _____, _____, _____, 89, 99

6. Add 10 to these numbers. Notice: you only change how many tens there are.

a. 61 + 10 = _____

   15 + 10 = _____

b. 34 + 10 = _____

   82 + 10 = _____

c. 69 + 10 = _____

   55 + 10 = _____

7. Count by fives. Use the chart to help. You can also say this aloud with your teacher.

a. 0, 5, _____, _____, _____, _____, _____, _____, _____, _____

b. 1, 6, _____, _____, _____, _____, _____, _____, _____, _____

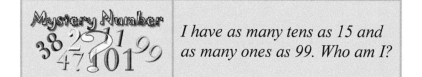

Mystery Number

I have as many tens as 15 and as many ones as 99. Who am I?

Try your own coloring patterns on this page!

1	2	3	4	5	6	7	8	9	10
11	12	13	14	15	16	17	18	19	20
21	22	23	24	25	26	27	28	29	30
31	32	33	34	35	36	37	38	39	40
41	42	43	44	45	46	47	48	49	50
51	52	53	54	55	56	57	58	59	60
61	62	63	64	65	66	67	68	69	70
71	72	73	74	75	76	77	78	79	80
81	82	83	84	85	86	87	88	89	90
91	92	93	94	95	96	97	98	99	100

1	2	3	4	5	6	7	8	9	10
11	12	13	14	15	16	17	18	19	20
21	22	23	24	25	26	27	28	29	30
31	32	33	34	35	36	37	38	39	40
41	42	43	44	45	46	47	48	49	50
51	52	53	54	55	56	57	58	59	60
61	62	63	64	65	66	67	68	69	70
71	72	73	74	75	76	77	78	79	80
81	82	83	84	85	86	87	88	89	90
91	92	93	94	95	96	97	98	99	100

Bar Graphs

This is a **bar graph**. Read it this way: look at the TOP of each column (bar), and look towards the left. How high does the top of the bar reach? Read the number.

Look at the first bar, for short pencils. Where does the top of that bar reach?

It reaches to 7. So, Henry has 7 short pencils.

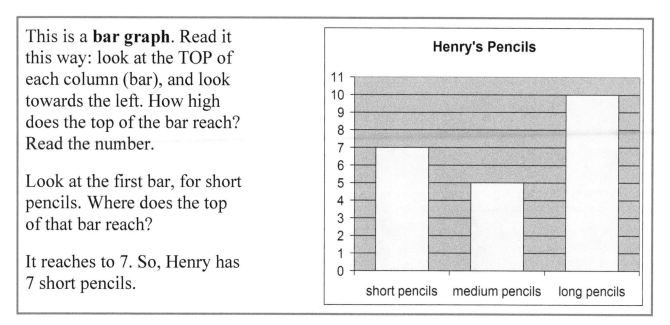

1. **a.** How many medium pencils does Henry have?

   b. How many long pencils does Henry have?

   c. How many short and medium pencils does Henry have in total?

   d. How many more long pencils does he have than short ones?

2. Here, the bar for first grade students reaches two little lines past 20. That's 22 students.

   a. How many students are in 2nd grade?

   b. How many students are in 3rd grade?

   c. How many students are in 4th grade?

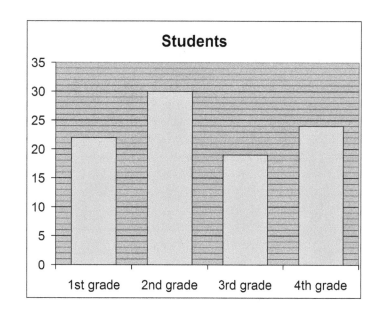

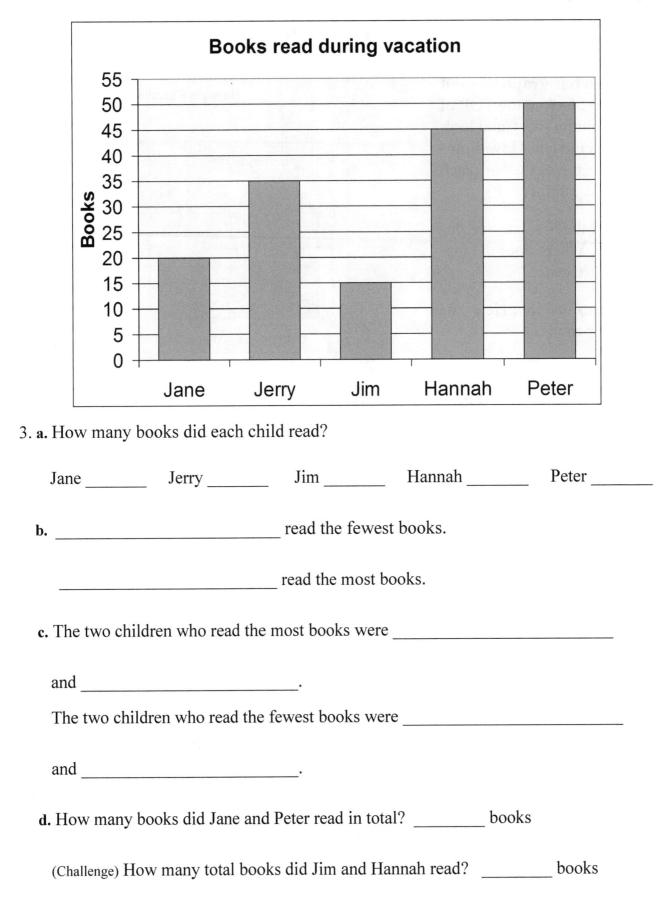

3. **a.** How many books did each child read?

Jane _____   Jerry _____   Jim _____   Hannah _____   Peter _____

b. _____ read the fewest books.

_____ read the most books.

c. The two children who read the most books were _____

and _____.

The two children who read the fewest books were _____

and _____.

d. How many books did Jane and Peter read in total? _____ books

(Challenge) How many total books did Jim and Hannah read? _____ books

Tally Marks

1. **Tally marks.** Tally marks are counting marks. When people count they make one tally mark for each thing they count. For one item or thing, draw one tally mark as " | ". The fifth tally mark is drawn across the four others like " ⊬⊬⊤ ".

Write the number that matches the tally.

⊬⊬⊤ | **a.** _____	⊬⊬⊤ ⊬⊬⊤ || **b.** _____	⊬⊬⊤ ⊬⊬⊤ ⊬⊬⊤ |||| **c.** _____	⊬⊬⊤ ⊬⊬⊤ ⊬⊬⊤ ⊬⊬⊤ ||| **d.** _____

2. Draw tally marks for these numbers.

a. 7	**b.** 14
c. 16	**d.** 32
e. 41	**f.** 28

3. Count the fish. Use tally marks to keep track. Mark each fish you count and make a tally mark for it. That way you won't count the same fish twice. Then write the number under "Count".

	Tally Marks	Count
Red		
Blue		
Yellow		

4. Count the pencils in each group. Use tally marks to keep track. Mark each pencil as you count it, and make a tally mark in the box. That way you won't count the same pencil twice.

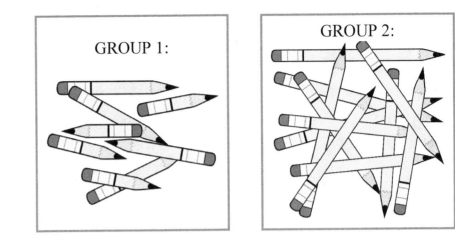

GROUP 1:

GROUP 2:

	Tally Marks	Count
Group 1		
Group 2		

5. Do the tally marks show the same counts that the bar graph does? If not, correct the tally.

	Tally Marks
Blue	⊞⊞ ⊞⊞ ⊞⊞ ⊞⊞ ⊞⊞ ‖
Black	⊞⊞ ⊞⊞
White	⊞⊞ ⊞⊞ ⊞⊞ ⊞⊞ ‖
Red	⊞⊞ ⊞⊞ ⊞⊞

Cars That Kyle Saw

6. (Optional) Tally marks are most useful for counting things that are happening rather slowly, for example, birds that fly into the yard. For this project, count something using tally marks. For example, you could go outside and count how many red and how many grey cars you see pass by your house in 20 minutes.

	Tally Marks	Count
Group 1		
Group 2		

Review Chapter 3

1. Name the numbers using numbers and words.

    a. 1 ten 5 ones   _15_   _____

    b. 6 tens 7 ones   _____   _____

    c. 4 tens 0 ones   _____   _____

    d. 10 tens 0 ones   _____   _____

    e. 5 tens 1 one   _____   _____

2. Fill in the numbers missing from the number lines.

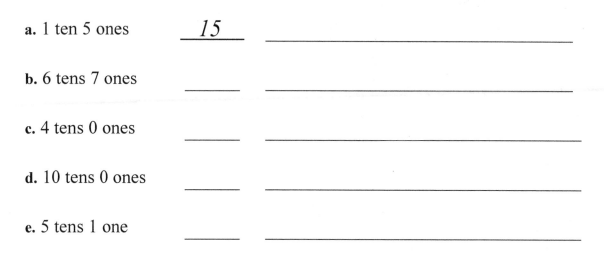

3. Circle the number that is *more*.

a.	**b.**	**c.**	**d.**	**e.**
78 87	22 25	56 57	68 80	101 11

4. Count. You can also say this aloud with your teacher.

97, 98, _____, _____, _____, _____, _____,

_____, _____, _____, _____, _____, _____.

5. Break the numbers into their tens and ones.

a. $45 = 40 + 5$	**b.** $25 = \_\_\_\_ + \_\_\_\_$	**c.** $78 = \_\_\_\_ + \_\_\_\_$
$68 = \_\_\_\_ + \_\_\_\_$	$54 = \_\_\_\_ + \_\_\_\_$	$91 = \_\_\_\_ + \_\_\_\_$

6. Build the numbers.

a. $50 + 7 = _____$	**b.** $8 + 10 = _____$	**c.** $90 + 6 = _____$
$20 + 0 = _____$	$9 + 70 = _____$	$9 + 60 = _____$

7. Put the numbers in order.

a. 57, 17, 75	**b.** 18, 48, 44, 41
$\_\_\_\_\_ < \_\_\_\_\_ < \_\_\_\_\_$	$\_\_\_\_\_ < \_\_\_\_\_ < \_\_\_\_\_ < \_\_\_\_\_$

8. Compare the expressions and write $<$, $>$ or $=$.

a. $56 \ \boxed{} \ 5 + 60$ **b.** $20 + 8 \ \boxed{} \ 33$ **c.** $60 + 5 \ \boxed{} \ 50 + 6$

d. $34 \ \boxed{} \ 30 + 6$ **e.** $4 + 90 \ \boxed{} \ 49$ **f.** $80 + 2 \ \boxed{} \ 70 + 9$

9. Skip-count. (You can say this aloud with your teacher.)

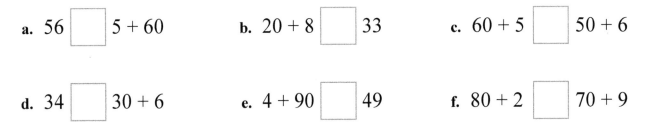

a. 13, 15, 17, _____, _____, _____, _____, _____, _____

b. _____, _____, _____, _____, _____, _____, 78, 88, 98

c. _____, _____, _____, _____, _____, 55, 60, 65, _____

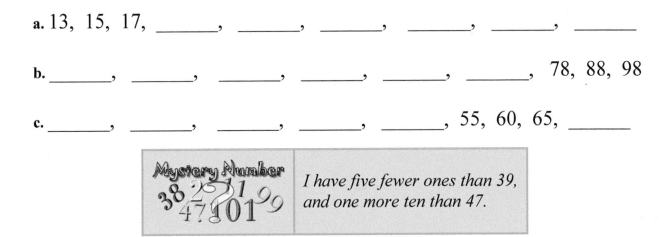

Mystery Number

I have five fewer ones than 39, and one more ten than 47.

CPSIA information can be obtained
at www.ICGtesting.com
Printed in the USA
LVHW061132061020
668071LV00013B/503